I0064934

Boundary Elements and other Mesh Reduction Methods XLVI

WITPRESS

WIT Press publishes leading books in Science and Technology.
Visit our website for the current list of titles.
www.witpress.com

WITeLibrary

Home of the Transactions of the Wessex Institute.
Papers published in this volume are archived in the WIT elibrary in volume 135 of WIT
Transactions on Engineering Sciences (ISSN 1743-3533).
The WIT electronic-library provides the international scientific community with immediate
and permanent access to individual papers presented at WIT conferences.
http://library.witpress.com

FORTY SIXTH INTERNATIONAL CONFERENCE ON
BOUNDARY ELEMENTS AND OTHER MESH REDUCTION METHODS

BEM/MRM 46

CONFERENCE CHAIRMEN

Alexander H. D. Cheng
University of Mississippi, USA
Member of WIT Board of Directors

Eduardo Divo
Embry-Riddle Aeronautical University, USA

Alain J. Kassab
University of Central Florida, USA

INTERNATIONAL SCIENTIFIC ADVISORY COMMITTEE

Andre Buchau
Weiqiu Chen
Jeng-Tzong Chen
Joao Batista de Paiva
Petia Dineva
Hrvoje Dodig
Chunying Dong
Ney Dumont
Zhuojia Fu
Alexander Galybin
Xiao-Wei Gao
Luis Godinho
Yiu Chung Hon
Andreas Karageorghis
John Katsikadelis
Edson Leonel
Daniel Lesnic
Leevan Ling
Yijun Liu
George Manolis
Liviu Marin

Toshiro Matsumoto
Leandro Palermo Jr
Ernian Pan
Andres Peratta
Dragan Poljak
Jure Ravnik
Joseph Rencis
Antonio Romero Ordonez
Bozidar Sarler
Martin Schanz
Vladimir Sladek
Elena Strelnikova
Stavros Syngellakis
Antonio Tadeu
Jon Trevelyan
Pihua Wen
Luiz Wrobel
Zhenhan Yao
Wolf Yeigh
Jianming Zhang
Chuanzeng Zhang

ORGANISED BY

Wessex Institute, UK
Embry-Riddle Aeronautical University, USA
University of Central Florida, USA
University of Mississippi, USA

SPONSORED BY

WIT Transactions on Engineering Sciences

WIT Transactions

Wessex Institute
Ashurst Lodge, Ashurst
Southampton SO40 7AA, UK

We would like to express thanks to all the conference Chairs and members of the International Scientific Advisory Committees for their efforts during the 2023 conference season.

Conference Chairs

Alexander Cheng
University of Mississippi, USA
(Member of WIT Board of Directors)

Eduardo Divo
Embry-Riddle Aeronautical University, USA

Alain Kassab
University of Central Florida, USA

Stavros Syngellakis
Wessex Institute, UK
(Member of WIT Board of Directors)

International Scientific Advisory Committee Members 2023

Maria Isabel Abreu Polytechnic Institute of Bragança, Portugal

Hussain Al-Kayiem University of Technology, Iraq

Joao Batista de Paiva University of São Paulo, Brasil

Andre Buchau University of Stuttgart, Germany

Weiqiu Chen Zhejiang University, China

Jeng-Tzong Chen National Taiwan Ocean University, Taiwan

Carmen Diaz Lopez University of Seville, Spain

Petia Dineva Bulgarian Academy of Sciences, Bulgaria

Hrvoje Dodig University of Split, Croatia

Chunying Dong Beijing Insitute of Technology, China

Ney Dumont Pontifical Catholic University of Rio de Janeiro, Brazil

Zhuojia Fu Hohai University, China

Alexander Galybin Schmidt Institute of Physics of the Earth, Russia

Xiao-Wei Gao Dalian University of Technology, China

Nasir Anka Garba Federal University Gusau, Nigeria

Emanuele Giorgi Tecnologico de Monterrey, Mexico

Luis Godinho University of Coimbra, Portugal

Yiu Chung Hon City University of Hong Kong, Hong Kong

Erik Jarlsby Eureka Energy Partners, Norway

Pushpa Jha Sant Longowal Institute of Engineering & Technology, India

Andreas Karageorghis University of Cyprus, Cyprus

John Katsikadelis National Technical University of Athens, Greece

Mikhail Kozhevnikov Ural Federal University, Russia

Edson Leonel University of São Paulo, Brazil

Daniel Lesnic University of Leeds, UK

Leevan Ling Hong Kong Baptist University, Hong Kong

Yijun Liu Southern University of Science and Technology, China

Elena Magaril Ural Federal University, Russia

Simone Maggiore Ricerca sul Sistema Energetico-RSE SpA, Italy

Nader Mahinpey University of Calgary, Canada

Robert Mahler University of Idaho, USA

George Manolis Aristotle University of Thessaloniki, Greece

Liviu Marin University of Bucharest, Romania

Liliana Marquez-Benavides Universidad Michoacana de San Nicolas de Hidalgo,Mexico

Toshiro Matsumoto Nagoya University, Japan

Felix Nieto University of A Coruña, Spain

Leandro Palermo Jr State University of Campinas, Brazil

Ernian Pan National Yang Ming Chiao Tung University, Taiwan

Deborah Panepinto Turin Polytechnic, Italy

Andres Peratta Beasy, UK

Dragan Poljak University of Split, Croatia

Elena Cristina Rada Insubria University of Varese, Italy

Marco Ragazzi University of Trento, Italy

Jure Ravnik University of Maribor, Slovenia

Joseph Rencis The University of Texas Permian Basin, USA

Antonio Romero Ordonez University of Sevilla, Spain

Bozidar Sarler University of Ljubljana, Slovenia

Martin Schanz Graz University of Technology, Austria

Antonio Serrano-Jimenez University of Granada, Spain

Nuno Simoes University of Coimbra, Portugal

Vladimir Sladek Slovak Academy of Sciences, Slovakia

Elena Strelnikova National Academy of Sciences of Ukraine, Ukraine

Antonio Tadeu University of Coimbra, Portugal

Jon Trevelyan University of Durham, UK
Jaap Vleugel Delft University of
    Technology, Netherlands
Peter Vorobieff University of New Mexico,
    USA
Pihua Wen Queen Mary University of
    London, UK
Luiz Wrobel Brunel University, UK
Zhenhan Yao Tsinghua University, China
Wolf Yeigh University of Washington,
    USA
Jianming Zhang Hunan University, China
Chuanzeng Zhang University of Siegen,
    Germany

Boundary Elements and other Mesh Reduction Methods XLVI

Editors

Alexander H. D. Cheng
University of Mississippi, USA
Member of WIT Board of Directors

Eduardo Divo
Embry-Riddle Aeronautical University, USA

Alain J. Kassab
University of Central Florida, USA

WITPRESS Southampton, Boston

Editors:

Alexander H. D. Cheng
University of Mississippi, USA
Member of WIT Board of Directors

Eduardo Divo
Embry Riddle Aeronautical University, USA

Alain J. Kassab
University of Central Florida, USA

Published by

WIT Press
Ashurst Lodge, Ashurst, Southampton, SO40 7AA, UK
Tel: 44 (0) 238 029 3223; Fax: 44 (0) 238 029 2853
E-Mail: witpress@witpress.com
http://www.witpress.com

For USA, Canada and Mexico

Computational Mechanics International Inc
25 Bridge Street, Billerica, MA 01821, USA
Tel: 978 667 5841; Fax: 978 667 7582
E-Mail: infousa@witpress.com
http://www.witpress.com

British Library Cataloguing-in-Publication Data

A Catalogue record for this book is available
from the British Library

ISBN: 978-1-78466-485-5
eISBN: 978-1-78466-486-2
ISSN: (print) 1746-4471
ISSN: (on-line) 1743-3533

The texts of the papers in this volume were set individually by the authors or under their supervision. Only minor corrections to the text may have been carried out by the publisher.

No responsibility is assumed by the Publisher, the Editors and Authors for any injury and/or damage to persons or property as a matter of products liability, negligence or otherwise, or from any use or operation of any methods, products, instructions or ideas contained in the material herein. The Publisher does not necessarily endorse the ideas held, or views expressed by the Editors or Authors of the material contained in its publications.

© WIT Press 2024

Open Access: All of the papers published in this volume are freely available, without charge, for users to read, download, copy, distribute, print, search, link to the full text, or use for any other lawful purpose, without asking prior permission from the publisher or the author as long as the author/copyright holder is attributed. This is in accordance with the BOAI definition of open access.

Creative Commons content: The CC BY 4.0 licence allows users to copy, distribute and transmit a paper, and adapt the article as long as the author is attributed. The CC BY licence permits commercial and non-commercial reuse.

Preface

This volume contains papers selected from the 46th International Conference on Boundary Elements and Other Mesh Reduction Methods (BEM/MRM 46). The conference series was founded by Professor Carlos Brebbia in 1978, with its first meeting held in Southampton, UK. For the next 46 years, scientists and engineers have used this gathering to exchange the progresses made in the field. The continued success of the meeting is a result of the strength of the research on boundary elements and mesh reduction techniques being carried out all over the world.

In the year 2020, the world and the conference faced a great challenge – the Covid-19 pandemic. All in-person scientific gatherings ceased. The BEM/MRM conferences were able to continue in a different format – both the BEM/MRM 43 and 44 were conducted online. The BEM/MRM 45 originally scheduled in an European location faced another challenge, the war in Ukraine. Hence, the 45th edition of the conference was also held online.

We were pleased that after three years of absence, the BEM/MRM 46 was held in person in Lisbon, Portugal, 3–5 October 2023. Colleagues and new researchers gathered for lively academic discussions and happy social events. We hope that we can continue to meet in this format in future years.

This volume collects some of the papers presented at the conference. The Editor would like to thank all authors for the quality of their papers and other colleagues for their help in reviewing the material.

The Editors
2023

Contents

Section 5: Fluid flow modelling

SECTION 1
ELECTRICAL ENGINEERING
AND ELECTROMAGNETICS

BOUNDARY ELEMENT BASED DOSIMETRY METHODS FOR THE ASSESSMENT OF HUMAN EXPOSURE TO RADIATION FROM 5G MOBILE SYSTEMS

DRAGAN POLJAK
FESB, University of Split, Croatia

ABSTRACT

Exposure of humans to 5G mobile communication systems may result in a local surface temperature elevation, i.e. may cause heating skin, ears and eyes. For the frequencies less than transition frequency of 6 GHz the specific absorption rate (SAR) is used to quantify the volume heating. However, according to recently published ICNIRP 2020 safety guidelines, the surface heating above 6 GHz is quantified by absorbed power density (S_{ab}). Furthermore, an alternative dosimetric quantity referred to as transmitted power density (TPD), is also used for internal dosimetry above 6 GHz. This paper aims to review some recently developed internal dosimetry methods based on the use of Galerkin–Bubnov indirect boundary element method for the assessment of human exposure to electromagnetic fields generated by 5G mobile systems. Different tissue models have been used in the paper. Some illustrative computational results have been presented.

Keywords: human exposure, 5G mobile communication systems, electromagnetic-thermal dosimetry, absorbed power density, transmitted power density, boundary element analysis.

1 INTRODUCTION

Local temperature increase at the human body surface is a well-established effect due to exposure to mobile communications systems of fifth generation (5G) [1], [2].

This surface heating is quantified by absorbed power density (S_{ab}) above transition frequency of 6 GHz according to IEEE 2019 and ICNIRP 2020 guidelines [1], [2]. An alternative quantity transmitted power density (*TPD*) can be also used [3], [4]. Note that specific absorption rate (*SAR*) is preferable quantity for frequencies below 6GHz.

This work aims to review recent work of the author on the assessment of S_{ab} and *TPD* in multi-layered tissue exposed to dipole antenna located horizontally to the interface by using boundary element method (BEM).

The assessment of S_{ab}/*TPD* is carried out in three steps:

- The first step in the assessment procedure is to determine the current distribution along the transmitting antenna using the Galerkin–Bubnov indirect boundary element method (GB-IBEM) to numerically handle the Pocklington integro-differential equation (IDE). The influence of the air-tissue interface is taken into account via the corresponding reflection/transmission coefficient.
- The second step is the evaluation of the electric and magnetic field, respectively, generated by the antenna. Provided the current distribution along the antenna is calculated the irradiated fields are determined by numerically calculating the corresponding field integrals by means of BEM formalism.
- The third step is to calculate S_{ab}/*TPD* by integrating radiated electric and magnetic fields according to the definition of S_{ab}/*TPD*.

More mathematical details pertaining to the formulation and related analytical/numerical procedures used to compute S_{ab} /*TPD* have been reported elsewhere, for example, in Poljak et al. [5], [6].

WIT Transactions on Engineering Sciences, Vol 135, © 2023 WIT Press
www.witpress.com, ISSN 1743-3533 (on-line)
doi:10.2495/BE460011

It is worth noting that according to ICNIRP 2020 guidelines S_{ab} is averaged over area of 4 cm^2 and 1 cm^2, respectively, depending on the operating frequency. The two-layer and three-layer tissue models are studied and some illustrative results for S_{ab} and *TPD* are given.

The present paper is organized as follows; Section 2 outlines the Pocklington IDE formulation for the assessment of the current distribution along the antenna, while the definitions for S_{ab} and *TPD* are given in Section 3. Numerical solution method is addressed in Section 4. The results are presented in Section 4, while last section pertains to some concluding remarks.

2 FORMULATION

Geometry of interest is transmitting dipole antenna of length L and radius a positioned parallel to the air-tissue interface (Fig. 1).

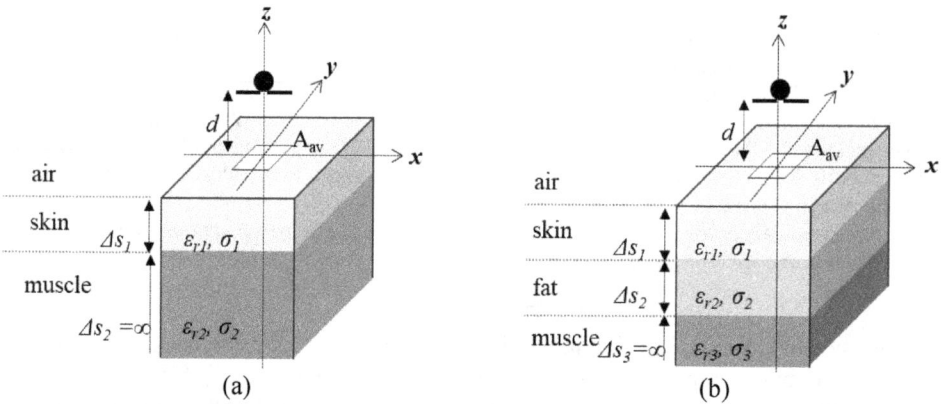

Figure 1:   Dipole antenna in front of planar multilayered tissue. (a) 2-layered (SM) model (skin and muscle layers); and (b) 3-layered (SFM) model (skin, fat and muscle).

Fig. 2 shows the propagation of the field radiated from the dipole antenna above the interface through the tissue.

The current distribution $I(x')$ flowing in the axis of straight wire antenna positioned parallel to a multilayered tissue is governed by the Pocklington integro-differential equation [7]

$$E_x^{exc} = j\omega \frac{\mu}{4\pi} \int_0^L I(x')g(x,x')dx' \; - \frac{1}{j4\pi\omega\varepsilon_0} \frac{\partial}{\partial x} \int_0^L \frac{\partial I(x')}{\partial x'} g(x,x')dx', \tag{1}$$

where E_x^{exc} is the electric field tangential to the wire due to an equivalent voltage source and $g(x,x')$ represents the total Green function:

$$g(x,x') = g_0(x,x') - \Gamma g_i(x,x'), \tag{2}$$

where $g_0(x,x')$ is the free space-Green function:

$$g_0(x,x') = \frac{e^{-jk_0 R_0}}{R_0}, \tag{3}$$

while $g_i(x,x')$ due to the image wire is:

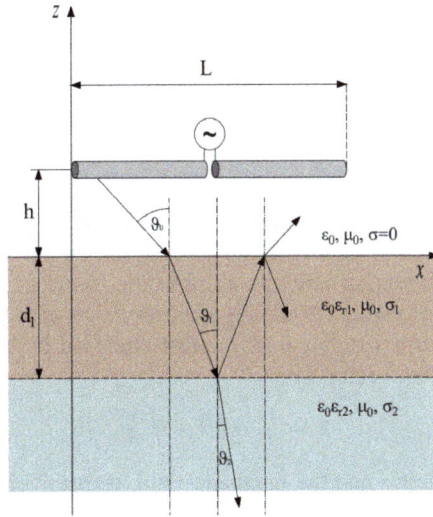

Figure 2: Horizontal dipole antenna in front of planar multilayered tissue.

$$g_i(x, x') = \frac{e^{-jk_0 R_i}}{R_i}, \tag{4}$$

where R_0 and R_i are the distances from the source and its image to the observation point of interest, respectively, while k is the phase constant of free space.

The reflection coefficient Γ accounts for the reflection from the interface. More details on the derivation of reflection/transmission coefficient for multilayer structure are available elsewhere, for example, in Poljak et al. [5], [6].

The key point in studying the wire antenna radiation is the assessment of the current distribution. Once the antenna current is calculated, or assumed, it is possible to determine the radiated fields.

Thus, the electric field components are [7]:

$$E_x \frac{1}{j4\pi\omega\varepsilon_{eff}} \left[-\int_0^L \frac{\partial I(x')}{\partial x'} \frac{\partial g(x,y,x')}{\partial x} dx' - \gamma^2 \int_{-L/2}^{L/2} I(x') g(x, x') dx' \right], \tag{5a}$$

$$E_y = \frac{1}{j4\pi\omega\varepsilon_{eff}} \int_0^L \frac{\partial I(x')}{\partial x'} \frac{\partial g(x,y,x')}{\partial y} dx', \tag{5b}$$

$$E_z = \frac{1}{j4\pi\omega\varepsilon_{eff}} \int_0^L \frac{\partial I(x')}{\partial x'} \frac{\partial g(x,y,x')}{\partial z} dx', \tag{5c}$$

while the magnetic field components are given by integrals [4]:

$$H_y = \frac{\mu}{4\pi} \int_0^L I(x') \frac{\partial g(x,y,z,x')}{\partial z} dx', \tag{6a}$$

$$H_z = -\frac{\mu}{4\pi} \int_0^L I(x') \frac{\partial g(x,y,z,x')}{\partial y} dx'. \tag{6b}$$

Calculating the electric and magnetic fields in the tissue the dosimetric quantities valid in GHz frequency range (above 6 GHz) can be determined.

3 ABSORBED/TRANSMITTED POWER DENSITY

Dosimetry above 6 GHz requires calculation of absorbed power density (S_{ab}), or the alternative quantity transmitted power density (*TPD*). Definition of S_{ab} and *TPD*, respectively, is given in subsections below.

3.1 Absorbed power density

Widely adopted quantity for the internal dosimetry above 6 GHz according to relevant recently updated guidelines and standards [1], [2] is absorbed power density (S_{ab}) stemming from the generalized conservation law of energy in the electromagnetic field given in the form of Poynting theorem [5] which describes a conservation of the energy stored in electric and magnetic fields within a volume V of interest enclosed by an area A. For the time-harmonic quantities Poynting vector S is

$$\vec{S} = \tfrac{1}{2} \, Re\big(\vec{E} \times \vec{H}^{*}\big), \tag{7}$$

where E and H are the electric and the magnetic field, respectively, due to some source of radiation.

Absorbed pover density S_{ab} adopted in ICNIRP 2020 is defined as [1]

$$S_{ab} = \tfrac{1}{2A_{av}} \int_{A_{av}} Re\big(\vec{E} \times \vec{H}^{*}\big) d\vec{A}, \tag{8}$$

where A_{av} is the averaging area, as depicted in Fig. 1.

The choice of averaging area has been addressed in many papers. For example, a circular area of 1 cm^2 has been used in Funahashi et al. [8], while some studies claim that the choice of averaging area of 2 or 4 cm^2 can be correlated with the average mass of 10 g for local SAR [9]. Eventually, Finally, the use of averaging areas of 4 cm^2 and 1 cm^2 has been documented in IEEE [1].

The effect of actual averaging area on *IPD* and S_{ab} due to the Hertz dipole radiation in free space and S_{ab} in the presence of a lossy half-space is available in Poljak and Dorić [10], [11]. More realistic case of dipole antenna above a lossy medium has been recently addressed in Poljak et al. [12].

3.2 Transmitted power density

The transmitted power density (*TPD*) is defined in terms of following integral [6]

$$TPD = \tfrac{1}{2} \int_{0}^{r} \sigma |E(\vec{r})|^{2} dr, \tag{9}$$

where r is the variable perpendicular to the body surface and point $r = 0$ pertains to the air-body interface, and σ stands for a tissue conductivity.

4 NUMERICAL SOLUTION

Numerical solution of Pocklington eqn (1) is carried out using GB-IBEM [1] which was documented elsewhere, for example, Poljak [13] and Poljak and El Khamlichi Drissi [14]. The procedure is outlined below.

It is convenient to write (1) in an operator form

$$KI = Y, \tag{10}$$

where K is a linear operator, I is the unknown current to be found for a given excitation E.

GB-IBEM starts by expanding the unknown current $I(x)$ into finite sum of linearly independent basis functions $f_n(x)$ with unknown complex coefficients I_n, i.e.

$$I_n(x') = \sum_{n=1}^{N_g} I_n f_n(x'),$$ (11)

where N_g is the number of base functions.

Substituting eqn (11) into eqn (10) one has

$$KI \cong KI_n = \sum_{i=1}^{n} \alpha_i KN_i = Y_n = P_n(Y).$$ (12)

The residual R is defined as

$$R = \sum_{n=1}^{N_g} I_n K(f_n) - E.$$ (13)

In accordance to the inner product of functions in Hilbert function space the error R is weighted to zero with respect to certain weighting functions $\{W_j\}$, i.e.

$$\int_L R \cdot W_m^* dx = 0, \quad m = 1, 2, .., N_g,$$ (14)

where (*) stands for the complex conjugate.

As K is linear, after some mathematical manipulation and by choosing $W_m = f_m$, (the Galerkin–Bubnov procedure) one obtains the following system of equations

$$\sum_{n=1}^{N_g} I_n \int_L K(f_n) f_m \, dx = \int_L E f_m dx, \quad m = 1, 2, \dots, N_g.$$ (15)

Utilizing the weak formulation [13], [14] and performing the integration by parts one obtains

$$\sum_{n=1}^{N_g} I_n \left[\int_{-L}^{L} \int_{-L}^{L} \frac{df_m(x)}{dx} \frac{df_n(x')}{dx'} g(x,x') dx' dx + k^2 \int_{-L}^{L} \int_{-L}^{L} f_m(x) f_n(x') g(x,x') dx' dx \right] =$$
$$-j4\pi\omega\varepsilon \int_{-L}^{L} E_x^{inc}(x) f_m(x) dx, \quad m = 1, 2, \dots, N_g$$ (16)

Eqn (16) is the weak Galerkin–Bubnov formulation of IDE (1) now requiring the basis and weight functions from the class of order-one differentiable functions.

Within the framework of the weak formulation boundary conditions are easily incorporated into the global matrix.

Applying the GB-IBEM algorithm and discretizing the wire leads to the global system of equations in the form of generalized Ohm's law

$$\sum_{i=1}^{M} [Z]_{ji} \{I\}_i = \{V\}_j, \quad j = 1, 2, \dots, M$$ (17)

where M is the total number of wire segments, $[Z]_{ji}$ is the mutual impedance matrix representing the interaction of the ith source to the jth observation segment, respectively and $\{V\}_j$ is the voltage vector for jth observation segment.

As functions $f(x)$ are required to be once differentiable a convenient choice for the shape of functions over the elements is the family of Lagrange's polynomials:

$$f_1(x) = \frac{x_2 - x}{\Delta x}, \, f_2(x) = \frac{x - x_1}{\Delta x}, \tag{18}$$

where x_1 and x_2 are the coordinates of the segment nodes and $\Delta x = x_2 - x_1$ is the segment length.

Now matrix $[Z]_{ji}$ and vector $\{V\}_j$ are given by

$$[Z]_{ji} = \int_{\Delta l_j} \int_{\Delta l_i} \begin{bmatrix} \dfrac{df_1(x)}{dx}\dfrac{df_1(x')}{dx'} & \dfrac{df_1(x)}{dx}\dfrac{df_2(x')}{dx'} \\ \dfrac{df_2(x)}{dx}\dfrac{df_1(x')}{dx'} & \dfrac{df_2(x)}{dx}\dfrac{df_2(x')}{dx'} \end{bmatrix} g(x, x') dx' dx$$

$$+ k^2 \int_{\Delta l_j} \int_{\Delta l_i} \begin{bmatrix} f_1(x)f_1(x') & f_1(x)f_2(x') \\ f_2(x)f_1(x') & f_2(x)f_2(x') \end{bmatrix} g(x, x') dx' dx$$

$$= \frac{1}{\Delta x^2}\frac{df_1(x')}{dx'} \int_{x_1}^{x_2} \int_{x_1}^{x_2} \begin{bmatrix} 1 & -1 \\ -1 & 1 \end{bmatrix} g_0(x, x') dx' dx$$

$$+ \frac{k^2}{\Delta x^2} \int_{x_1}^{x_2} \int_{x_1}^{x_2} \begin{bmatrix} (x_2 - x)(x_2 - x') & (x_2 - x)(x' - x_1) \\ (x - x_1)(x_2 - x') & (x - x_1)(x' - x_1) \end{bmatrix} g_0(x, x') \, dx' dx, \tag{19}$$

$$\{V\}_j = -j4\pi\omega\varepsilon \int_{\Delta l_j} E_x^{inc}(x) \begin{bmatrix} f_1(x) \\ f_2(x) \end{bmatrix} dx = -\frac{j4\pi\omega\varepsilon}{\Delta x} \int_{x_1}^{x_2} E_x^{inc}(x) \begin{bmatrix} (x_2 - x) \\ (x - x_1) \end{bmatrix} dx, \tag{20}$$

where Δl_i, Δl_j are the widths of ith and jth segments.

The voltage vector can be computed analytically for the case of delta-function source (antenna mode), or the plane wave excitation (scatterer mode). In the antenna mode the voltage vector vanishes outside the feed gap area.

The excitation field is

$$E_x^{inc}(x) = \frac{V_g}{\Delta l_g} \tag{21}$$

where V_g is the feed voltage and $\Delta l_g = \Delta x$ (for convenience) is the feed-gap width.

Using the linear shape functions yields

$$\{V\}_j = -\frac{j4\pi\omega\varepsilon}{\Delta l_g} \int_{x_1 = -\frac{\Delta l_g}{2}}^{x_2 = \frac{\Delta l_g}{2}} \frac{V_g}{\Delta l_g} \begin{bmatrix} (x_2 - x) \\ (x - x_1) \end{bmatrix} dx = -j2\pi\omega\varepsilon V_g \begin{pmatrix} 1 \\ 1 \end{pmatrix}. \tag{22}$$

In the scattering mode for the simple case of normal incidence the wire is excited by the plane wave, i.e.:

$$E_x^{inc}(x) = E_0, \tag{23}$$

and the voltage vector is given by:

$$\{V\}_j = -\frac{j4\pi\omega\varepsilon}{\Delta l} \int_{x_1 = -\frac{\Delta l}{2}}^{x_2 = \frac{\Delta l}{2}} E_0 \begin{bmatrix} (x_2 - x) \\ (x - x_1) \end{bmatrix} dx = -j2\pi\omega\varepsilon E_0 \Delta l - \begin{pmatrix} 1 \\ 1 \end{pmatrix}. \tag{24}$$

More details on the mathematical description of the method can be found elsewhere, e.g. in Poljak [13].

5 NUMERICAL RESULTS

The results for S_{ab} and *TPD* are presented in this section. The results obtained via numerical approach are compared against analytical results [5], [6]. Frequency dependent electrical parameters for the body tissues are available in ITIS Foundation [15].

Table 1: Frequency dependent parameters of body tissues [15].

f (GHz)	Skin		Fat		Muscle	
	ε_r	σ (S/m)	ε_r	σ (S/m)	ε_r	σ (S/m)
6	34.9	3.89	9.80	0.872	48.2	5.20
10	31.3	8.01	8.80	1.71	42.8	10.6

The first set of results deals with the results for S_{ab} vs distance from the interface for different antenna lengths ($L = \lambda/2$, $L = \lambda/4$ and $\lambda/10$). The operating frequency of the transmitting antenna is $f = 10$ GHz. The control surface is $A = 1$ cm^2 and 4 cm^2, respectively. The results obtained via GB-IBEM are compared to the results obtained via assumed sinusoidal current distribution (Fig. 3).

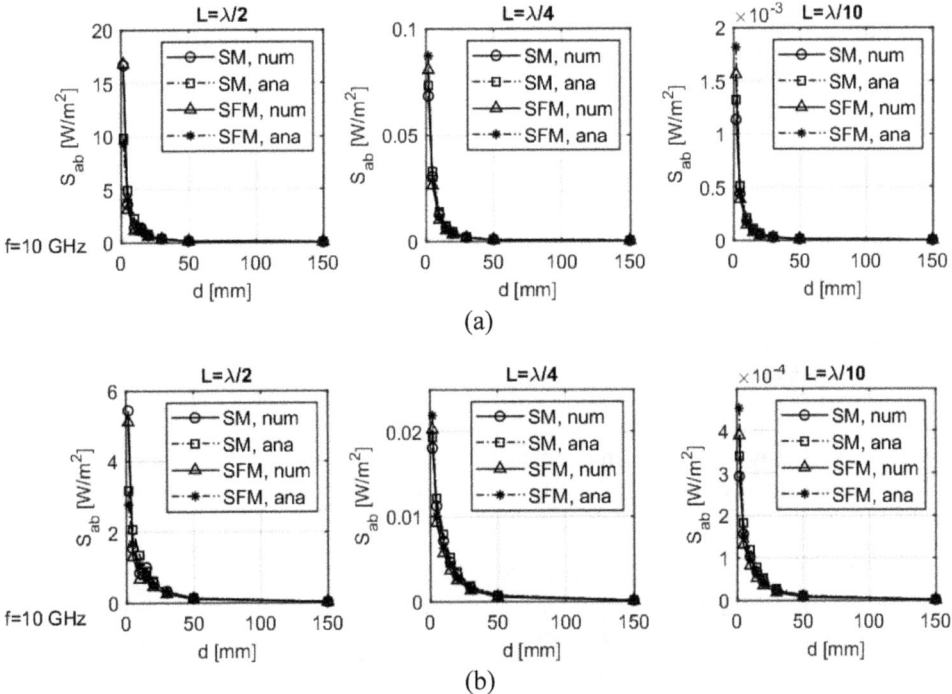

Figure 3: S_{ab} vs antenna-body distance at $f = 10$ GHz for different wire lengths and control surface. (a) $A = 1$ cm^2; and (b) $A = 4$ cm^2.

Discrepancies in S_{ab} values obtained via numerical/analytical approach are approximately, 10% and 20% for wire lengths $L = \lambda/4$ and $\lambda/10$. Also, the discrepancies between SM and SFM models decrease as frequency and antenna-body distance increase.

Fig. 4 pertains to the assessment of *TPD* vs antenna-body distance for different antenna lengths for operating frequency f=10GHz.

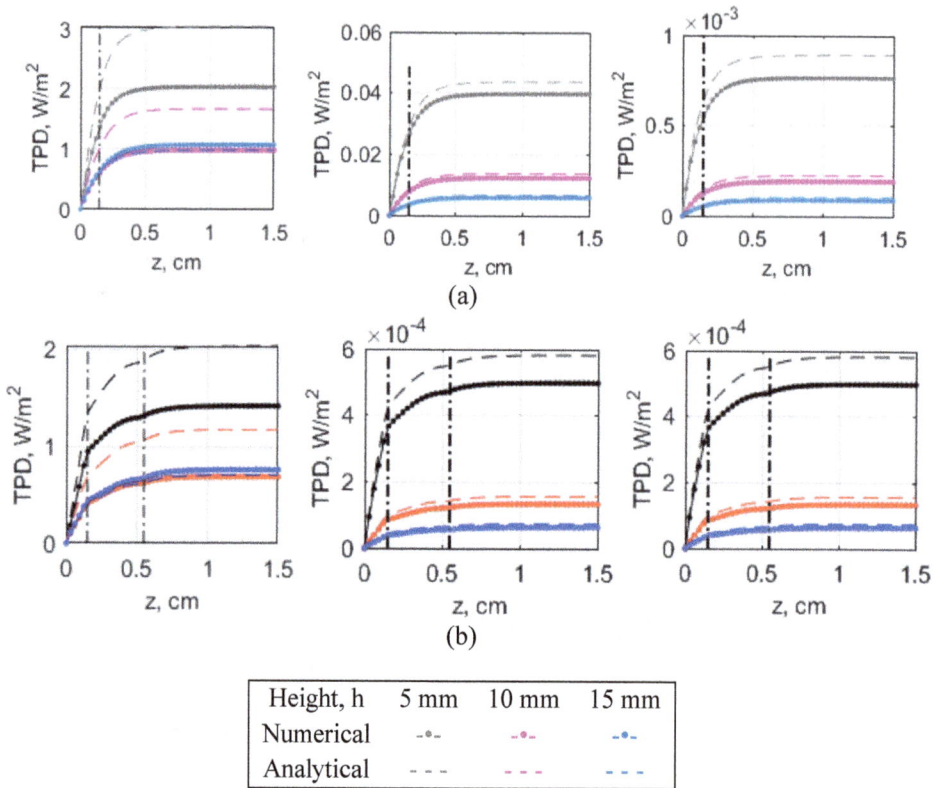

Height, h	5 mm	10 mm	15 mm
Numerical	–•–	–•–	–•–
Analytical	– – –	– – –	– – –

Figure 4: *TPD* versus tissue depth for different antenna lengths ($L = \lambda/2$, $L = \lambda/4$ and $L = \lambda/10$) above multilayered tissue.

TPD obviously grows until saturation rapidly. The discrepancy between analytical/numerical results is highest for $L = \lambda/2$ and increases as antenna-body distance decreases.

6 CONCLUSION

The paper reviews BEM based approaches to determine the dosimetry quantities in GHz frequency range (millimetre waves); absorbed power density (S_{ab}) and transmitted power density (*TPD*) being of interest for 5G mobile systems. Planar two-layer and three-layer tissue models exposed to dipole radiation have been addressed. Some illustrative analytical/numerical results for S_{ab} and *TPD* are given. The obtained results show that S_{ab} decreases rapidly with the increase of the antenna-body distance. The higher is the antenna length, the higher is S_{ab}. Analysing the results for *TPD* it is obvious that the highest differences between the analytical/numerical results are found for $\lambda/2$ dipole. The discrepancy decreases with the antenna length.

REFERENCES

[1] IEEE, IEEE Standard for Safety Levels with Respect to Human Exposure to Electric, Magnetic, and Electromagnetic Fields, 0 Hz to 300 GHz. IEEE Std C95.1-2019 (Revision of IEEE Std C95.1-2005/ Incorporates IEEE Std C95.1-2019/Cor 1-2019), pp. 1–312, 2019.

[2] ICNIRP, Guidelines for limiting exposure to electromagnetic fields (100 kHz to 300 GHz). *Health Physics*, **118**(5), pp. 483–524, 2020.

[3] Pfeifer, S. et al., Total field reconstruction in the near field using pseudo-vector E-field measurements. *IEEE Transactions on Electromagnetic Compatibility*, **61**(2), pp. 476–486, 2019.

[4] Hirata, A., Funahashi, D. & Kodera, S., Setting exposure guidelines and product safety standards for radio-frequency exposure at frequencies above 6 GHz: Brief review. *Annals of Telecommunications*, **74**, pp. 17–24, 2019.

[5] Poljak, D., Susnjara, A. & Kraljevic, L., Assessment of absorbed power density in multilayer planar model of human tissue. *Radiation Protection Dosimetry*, **199**(8–9), pp. 798–805, 2023.

[6] Poljak, D., Susnjara, A. & Fisic, A., Assessment of transmitted power density in the planar multilayer tissue model due to radiation from dipole antenna. *Journal of Communications Software and Systems*, **19**(1), pp. 39–51, 2023. DOI: 10.24138/jcomss-2022-0050.

[7] Poljak, D. et al., Absorbed power density in a multilayer tissue model due to radiation of dipole antenna at GHz frequency range. Part I: Theoretical background. *Proc. Splitech*, Split, Croatia, July, 2022.

[8] Funahashi, D. et al., Area-averaged transmitted power density at skin surface as metric to estimate surface temperature elevation. *IEEE Access*, **6**, pp. 77665–77674, 2018.

[9] Funahashi, D. et al., Averaging area of incident power density for human exposure from patch antenna arrays. *IEICE Transactions on Electronics*, **E101-C**(8), pp. 644–646, 2018.

[10] Poljak, D. & Dorić, V., On the concept of the transmitted field and transmitted power density for simplifed case of hertz dipole. *EMC Europe 2020*, Rome, Italy, 2020.

[11] Poljak, D. & Dorić, V., Assessment of transmitted power density due to hertz dipole radiation using the modified image theory approach. *28th International Conference on Software, Telecommunications and Computer Networks (SoftCOM)*, Split, Croatia, 2020.

[12] Poljak, D., Šušnjara, A. & Džolić, A., Assessment of transmitted power density due to radiation from dipole antenna of finite length. Part I: Theoretical background and current distribution. *29th International Conference on Software, Telecommunications and Computer Networks*, Hvar, Croatia, 2021.

[13] Poljak, D., *Advanced Modeling in Computational Electromagnetic Compatibility*, John Wiley: New York, 2007.

[14] Poljak, D. & El Khamlichi Drissi, K., *Computational Methods in Electromagnetic Compatibility, Antenna Theory Approach versus Transmission Line Models*, John Wiley: New York, 2018.

[15] ITIS Foundation, Copyright © 2010–2021 ITIS Foundation, June 2019. https://itis.swiss/virtual-population/tissue-properties/database/.

ANALYSIS OF PLANAR SKIN MODEL EXPOSED TO DIPOLE ANTENNA RADIATION FEATURING THE USE OF THE MOM-MOM APPROACH

MARIO CVETKOVIĆ & DRAGAN POLJAK
Faculty of Electrical Engineering, Mechanical Engineering and Naval Architecture, University of Split, Croatia

ABSTRACT

The paper deals with high frequency dosimetry analysis of a planar skin model exposed to a dipole antenna radiating in the GHz frequency range. The thin equivalent strip of a perfect electrically conducting (PEC) center-fed antenna is based on the electric field integral equation (EFIE) formulation in the frequency domain solved by means of method of moments (MoM), while the simple planar model for high frequency assessment is based on EFIE formulation for lossy homogeneous biological body solved by the same numerical approach. The electric field obtained using the EFIE-PEC part is subsequently utilized as the incident field in the EFIE-dielectric part of the proposed model. The numerical results for the surface antenna current, the equivalent surface current, and the induced electric field on the surface of the planar model are given, as well as a comparison of several field measures obtained at the averaging surface of planar models of varying thickness and width, respectively. The results could be found useful in the development of computational dosimetry models in the assessment of exposure of humans to electromagnetic fields in the GHz frequency range.

Keywords: planar skin model, dipole antenna radiation, electric field integral equation, method of moments, human exposure to EMF.

1 INTRODUCTION

The use of computational dosimetry models in the assessment of human exposure to high-frequency electromagnetic fields (EMFs) is considered essential as the experimental validation is very difficult to perform in this frequency range. The improvement in computational methods is important basis for the development of exposure standards related to the safe use of EMFs [1]. According to International Commission on Non-Ionizing Radiation Protection (ICNIRP) [2], and IEEE C95.1. standard [3], respectively, the basic restrictions or the dosimetric reference levels in the frequency range of 6–300 GHz, are set in terms of the absorbed power density (APD), while below 6 GHz the exposure limits are specified in terms of specific absorption rate (SAR). Calculation of both SAR and APD is carried out by post-processing the results from the electromagnetic dosimetry model, i.e., the induced electric field. The subsequent thermal dosimetry models utilises the SAR or APD as inputs in order to determine the resulting temperature rise in the model of exposed biological tissue.

Various research groups within IEEE International Committee on Electromagnetic Safety (ICES) Subcommittee 6 study the effects of different averaging schemes of dosimetric quantities [4], [5]. These include different schemes for spatial averaging of the APD above 6 GHz, with averaging area of 4 cm^2, and 1 cm^2, for frequencies above 30 GHz, to correlate with the local maximum temperature increase. The initial step in this analysis is related to comparison of the area-averaged APD using planar skin models. The study in Li et al. [5] determined the IPD and APD averaged over the skin surface in the range of 10 to 90 GHz, employing various numerical approaches using uniform body and antenna models. To this end, different research group employed commercial electromagnetic solvers such as CST, HFSS, and Sim4Life, as well as a numerical research codes mainly based on the FDTD. On the other hand, the use of integral equation based techniques such as the method of moments (MoM), the boundary element method (BEM), or hybrid methods is limited due to computational requirements related to frequency of the problem. The fully populated matrices in the GHz frequency range becomes prohibitively large to solve on typical desktop computers, unless GPU acceleration or parallel processing is employed [6].

This paper is on the use of a surface integral equation formulation and the method of moments (MoM) based approach to high frequency dosimetry [7]. A scenario with single $\lambda/2$ dipole antenna positioned in front of a planar body model is considered in this study. The antenna is modelled as a perfect electric conductor (PEC) radiating at 10 and 30 GHz. The antenna model is represented by a thin-strip that is formulated using the electric field integral equation (EFIE) for PEC. The EFIE formulation solved by means of MoM technique is utilized first to find the surface current distribution due to an applied voltage in the antenna feed [8]. From the determined antenna surface current distribution, the radiated electric field is found, representing the incident electric field in the subsequent step, where the EFIE formulation for the homogeneous penetrable scatterer (body) is used [9]. The numerical solution has been carried out using MoM, resulting in the induced electric field at the scatterer surface. Some illustrative examples for electric field induced on the planar model surface are given.

The outline of the manuscript is as follows. Following the introduction, the EFIE formulation for PEC object is briefly outlined, utilized in order to find the dipole antenna radiation. This is followed by a brief outline of the EFIE formulation for lossy dielectric biological body represented by the planar model. More details on the utilized planar models is given in the subsequent section including some remarks on the computational complexity regarding the use of said models at very high frequencies. The numerical results including the dipole antenna current distribution, the planar model equivalent current distribution, as well as the computed electric field are given in the next section, while concluding remarks are given in the final section.

2 MATHEMATICAL DETAILS

2.1 EFIE-PEC formulation for the dipole antenna

The frequency domain (FD) EFIE for perfect conductor is utilized for the dipole antenna model. The formulation is based on the application of equivalence theorem and appropriate boundary conditions at the antenna surface, leading to:

$$[-\vec{E}_1^{sca}(\vec{J})]_{tan} = [\vec{E}^{inc}]_{tan}, \tag{1}$$

where E^{inc} and E^{sca} denote the known incident field (related to voltage at antenna terminals) and the unknown scattered field, respectively. The scattered field can be expressed as:

$$\vec{E}^{sca} = -j\omega\vec{A} - \nabla\varphi, \tag{2}$$

where $\vec{A}$ and φ denote the vector magnetic and the scalar electric potential, respectively, defined as:

$$\vec{A}(\vec{r}) = \mu \iint_S \vec{J}(\vec{r}')G(\vec{r},\vec{r}')dS'; \quad \varphi(\vec{r}) = \frac{1}{\varepsilon} \iint_S \rho(\vec{r}'), G(\vec{r},\vec{r}')dS', \tag{3}$$

where $\vec{J}$ and ρ are surface current density and surface charge density, respectively, and $G(\vec{r},\vec{r}')$ denotes the Green's function for free space:

$$G(\vec{r},\vec{r}') = \frac{e^{-jkR}}{4\pi R}; \quad R = |\vec{r} - \vec{r}'|, \tag{4}$$

where R denotes the distance between observation point $\vec{r}$ and source point $\vec{r}\,'$, and k is free space wave number.

The surface charges from (3) can be expressed, through the continuity equation, as:

$$\rho(\vec{r}\,') = -\frac{\nabla'_S \cdot \vec{J}(\vec{r}\,')}{j\omega}. \tag{5}$$

Inserting (2) in (1), after some rearranging, results in:

$$\vec{E}^{inc} = j\omega\mu \iint_S \vec{J}(\vec{r}\,')G(\vec{r},\vec{r}\,')dS' - \frac{j}{\omega\varepsilon} \iint_S \nabla'_S \cdot \vec{J}(\vec{r}\,')\nabla G'(\vec{r},\vec{r}\,')dS', \tag{6}$$

where $\nabla'_S\cdot$ denotes the surface divergence operator.

The unknown surface current $\vec{J}$ are expanded by linear combination of basis functions such as the Rao–Wilton–Glisson (RWG) functions [8], as follows:

$$\vec{J}(\vec{r}) = \sum_{n=1}^{N} J_n \vec{f}_n(\vec{r}), \tag{7}$$

with J_n denoting the unknown coefficients and N denoting the number of triangles used to discretize the antenna surface S.

After testing with $\vec{t}_m(\vec{r})$, where $\vec{t}_m = \vec{f}_n$, followed by integration over surface S, the gradient from Green's function can be transfered to testing function, resulting in the integrals of the following form [10]:

$$
\begin{aligned}
A &= \iint_S \vec{t}_m(\vec{r}) \cdot \iint_{S'} \vec{f}_n(\vec{r}\,')G(\vec{r},\vec{r}\,')\,dS'\,dS, \\
B &= \iint_S \nabla_S \cdot \vec{t}_m(\vec{r}) \iint_{S'} \nabla'_S \cdot \vec{f}_n(\vec{r}\,')G(\vec{r},\vec{r}\,')\,dS'\,dS, \\
V &= \iint_S \vec{t}_m(\vec{r}) \cdot \vec{E}^{inc}\,dS.
\end{aligned}
\tag{8}
$$

The resulting linear equations set can be written in the matrix form as

$$[\mathbf{Z}]_{PEC} \cdot \{\mathbf{I}\}_{PEC} = \{\mathbf{V}\}_{PEC}, \tag{9}$$

where the matrix $\mathbf{Z}$ size is $N \times N$, and the size of source vector $\mathbf{V}$ is N. Solving (9) leads to vector $\mathbf{I}$ with unknown coefficients J_n. From these coefficients, the antenna surface current $\vec{J}$ can be determined, and afterwards, the electric field can be found using:

$$\vec{E}(\vec{r}) = -j\omega\mu \iint_S \vec{J}(\vec{r}\,')G(\vec{r},\vec{r}\,')\,dS' - \frac{j}{\omega\varepsilon} \iint_S \nabla'_S \cdot \vec{J}(\vec{r}\,')G(\vec{r},\vec{r}\,')\,dS'. \tag{10}$$

2.2 EFIE-dielectric formulation for the planar model

The FD-EFIE for the lossy dielectric is utilized for the planar body model. The formulation is based on the use of the equivalence theorem [11] and the pertinent boundary conditions at the body surface, leading to the following:

$$\left[-\vec{E}^{sca}_n(\vec{J},\vec{M})\right]_{tan} = \begin{cases} \left[\vec{E}^{inc}\right]_{tan}, & i = 1 \\ 0, & i = 2, \end{cases} \tag{11}$$

where E^{inc} represents the radiated field by a dipole antenna while E^{sca} is the scattered electric field.

The tangential (*tan*) component of E^{sca} can be written in terms of the equivalent current densities, namely $\vec{J}$ and $\vec{M}$, which are then expanded as a linear combination of basis functions. If the triangular elements are used to describe the body surface, $\vec{J}$ can be expanded using RWG basis functions [8], while $\vec{M}$ can be expanded by $\hat{n} \times$ RWG, as:

$$\vec{J}(\vec{r}) = \sum_{n=1}^{N} J_n \vec{f}_n(\vec{r}); \quad \vec{M}(\vec{r}) = \sum_{n=1}^{N} M_n \vec{g}_n(\vec{r}), \qquad (12)$$

where J_n and M_n represent the coefficients to be solved for, while N is the number of elements used to discretize the surface of the body.

The numerical solution of EFIE is carried out using an efficient scheme based on MoM reported in Cvetković et al. [9]. In the following, only basic steps are outlined. More details can be found in authors previous work, e.g., Cvetković et al. [12], [13].

Multiplying (11) by testing functions $\vec{f}_m$, with $\vec{f}_m = \vec{f}_n$, followed by integration over the complete body surface S, after some additional steps [9], the following integral equations set is obtained:

$$\sum_{n=1}^{N} \left(j\omega\mu_i A_{mn,i} + \frac{j}{\omega\varepsilon_i} B_{mn,i} \right) J_n + \sum_{n=1}^{N} \left(C_{mn,i} + D_{mn,i} \right) M_n = \begin{cases} V_m, & i = 1 \\ 0, & i = 2, \end{cases} \qquad (13)$$

where A_{mn}, B_{mn}, C_{mn}, and D_{mn} are the double surface integrals, while index $i = 1, 2$ denote the outside and inside region, with respect to body. The m and n labels, respectively, denote the triangles with source and observation points. The body's electrical properties are accounted via μ and ε parameters, namely, permeability and permittivity.

The integral equations set (13) can be expressed compactly in the matrix form as

$$[\mathbf{Z}] \cdot \{\mathbf{I}\} = \{\mathbf{V}\}, \qquad (14)$$

where the size of system matrix $\mathbf{Z}$ is $2N \times 2N$, while the size of source vector $\mathbf{V}$ is $2N$. It should be noted that matrix $\mathbf{Z}$ is fully populated [7]. The unknown coefficients J_n and M_n are determined by the solution of (14), and are used to calculate the equivalent surface electric and magnetic currents, $\vec{J}$ and $\vec{M}$, respectively.

The electric field inside and on the body surface is thus easily determined, from the equivalent surface currents, using the following:

$$\vec{E}_2(\vec{r}) = -j\omega\mu_2 \iint_S \vec{J}(\vec{r}')G_2(\vec{r},\vec{r}')\,dS' - \frac{j}{\omega\varepsilon_2} \iint_S \nabla'_S \cdot \vec{J}(\vec{r}')G_2(\vec{r},\vec{r}')\,dS' -$$
$$- \iint_S \vec{M}(\vec{r}') \times \nabla G_2(\vec{r},\vec{r}')\,dS'. \qquad (15)$$

Likewise, the magnetic field can be calculated. Both fields, determined at the body surface, can be latter utilized in the calculation of absorbed power density, using one of two definitions [4], [5]:

$$S_{ab_1}(\vec{r}) = \frac{1}{2A} \iint_A \int_{z_1}^{z_2} \sigma(\vec{r})|\vec{E}(\vec{r})|^2 \, dz \, dA, \qquad (16)$$

Table 1: Model dimensions, number of RWG elements N, number of tetrahedras (Tetra), matrix fill time T_{fill}, and memory allocation for system matrix $[Z]$.

Model	Size (cm^3)	N	Tetras	T_{fill} (days:h:min)	$[Z]$ (MB)
#1	1 cm × 1 cm × 0.1 cm	720	960	01:17	35 MB
#2	1 cm × 1 cm × 0.2 cm	840	1038	01:46	52 MB
#4	1 cm × 1 cm × 0.4 cm	1080	1698	02:55	94 MB
#5	1 cm × 1 cm × 0.5 cm	1200	2058	3:41	118 MB
#10	1 cm × 1 cm × 1 cm	1800	3182	9:14	271 MB
#4x1.2	1.2 cm × 1.2 cm × 0.4 cm	1440	1982	5:35	167 MB
#4x1.6	1.6 cm × 1.6 cm × 0.4 cm	2304	3495	17:10	421 MB
#4x1.8	1.8 cm × 1.8 cm × 0.4 cm	2808	4303	1d:03:09	621 MB
#4.x2.0	2 cm × 2 cm × 0.4 cm	3360	5268	1d:18:44	880 MB
#4x3.0	3 cm × 3 cm × 0.4 cm	6840	11718	10d:07:16	3.55 GB

where σ is the conductivity, or a more rigorous definition [5]:

$$S_{ab_2}(\vec{r}) = \frac{1}{2A} \iint_A \Re \left[\vec{E}(\vec{r}) \times \vec{H}^*(\vec{r}) \right] \cdot d\vec{A}, \tag{17}$$

where $\vec{E}$ and $\vec{H}$ represent the peak value of electric field and magnetic field, respectively, on the body surface, $\Re$ is the field real part and (*) denotes the complex conjugate. It should be noted that the control surface of area A should be sufficiently larger than the EM penetration depth δ [4].

3 PLANAR SKIN MODELS

Various exposure scenarios have been considered in Li et al. [5], including a single-layer skin model whose dimensions (in mm) were $W \times H \times L = 200 \times 200 \times 100$, and $W \times H \times L = 150 \times 150 \times 75$, at 10 GHz and 30 GHz, respectively. If the conventional MoM is used, the computational size of the problem at these frequencies becomes a limiting factor [6]. Table 1 shows computational requirements for the considered models.

Let us consider e.g., a full scale planar model at 10 GHz [5]. Taking into account that element size should be on the order of $\lambda/10$, at 10 GHz, this translates to 3 mm maximum triangle side, and 1 mm at 30 GHz. Tesselating the surface with structured mesh results in the smallest number of elements equal to 225.000 triangles. This in turn results in $N = 375.000$ RWG elements, i.e., a fully populated matrix whose size is $(2N)^2$. Using a fairly decent laptop computer, featuring 4 cores for parallel computation, the computation time in case of $N = 3360$ RWG elements, according to Table 1, takes around 42 hours.

We are thus interested to find out whether the use of the reduced planar models, as shown in Fig. 1, could still be utilized, facilitating the use of conventional MoM.

We consider averaging over 1 cm^2 square area, employing models with varying thickness and width, as shown in Fig. 1, whose parameters are given in Table 2.

The geometry of the problem is depicted in Fig. 2.

Several locations of dipole antenna are considered, i.e., distance $d = 5, 10, 15$ mm with respect to the planar body.

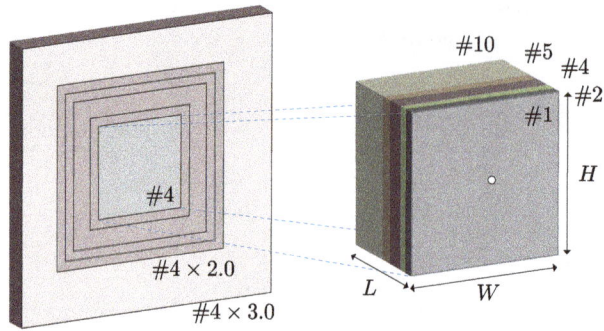

Figure 1: Used planar skin models. Models denoted $\#1 - \#10$ with different thickness L, while models $\#4 - \#4x3.0$ are with different front surface area. Dimensions $(W \times H \times L)$ are given in Table 1.

Table 2: Parameters of planar skin models.

	σ	ε_r	δ	$\lambda/10$
	(S/m)		(mm)	(mm)
10 GHz	8.84	32.41	3.79	3
30 GHz	27.31	16.63	0.85	1

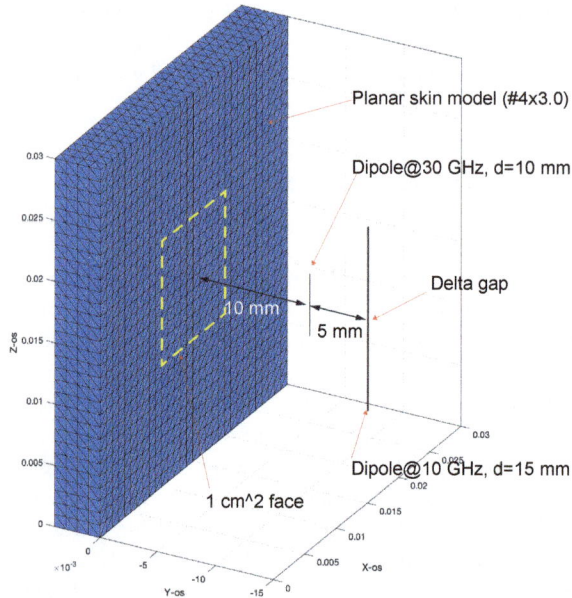

Figure 2: Geometry of the problem. Example of 30 GHz and 10 GHz halfwave dipoles, positioned 10 mm and 15 mm, respectively, in front of the planar model $\#4 \times 3.0$. Dotted square denotes 1 cm^2 area where averaging is carried out.

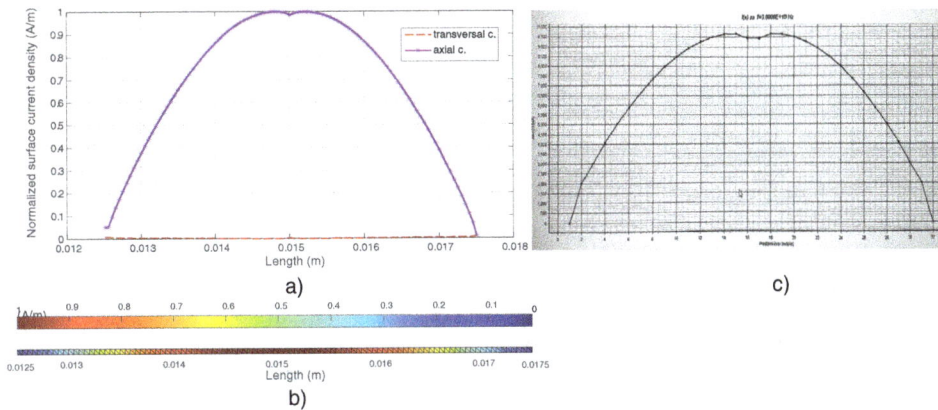

Figure 3: Example for 30 GHz halfwave dipole, dipole length $L = \lambda/2 = 5$ mm. (a) Distribution of transversal and axial component of surface current along dipole length; (b) Normalized current distribution along dipole represented by thin strip of triangular elements; and (c) Axial current obtained using SuZaNa numerical code [16].

4 NUMERICAL RESULTS

Some illustrative examples are presented in this section. The computational results are obtained using the EFIE formulation for PEC (antenna) and dielectric (body), respectively.

4.1 Dipole antenna surface current distribution

The antenna represented by thin strip of triangular elements is center-fed where the concept of delta gap is used [14]. The ideal voltage generator across a small gap is connected, assuming a gap of negligible width, thus easily implementable with RWG elements [15].

Both axial and transversal components of the antenna surface currents are depicted in Fig. 3(a).

Note the several orders of magnitude higher axial current component compared to the transversal one.

The comparison between EFIE-PEC formulation and the SuZaNa numerical code [16], as depicted in Figs 3(a) and (c), shows an excellent agreement.

4.2 Planar model equivalent surface currents

From the antenna current, the radiated field at any point can be calculated. If the radiated antenna field is used as the incident field in the subsequent EFIE-dielectric formulation, the equivalent electric and magnetic currents induced on the surface of planar body, can be determined, as depicted in Figs 4 and 5.

Figs 4 and 5 illustrates similar distribution of equivalent currents obtained using models of varying thickness and width, respectively.

4.3 Induced electric field

The induced electric field at the planar model surface can be calculated from the equivalent surface currents. An example of the numerical results for the electric field obtained at the surface of planar skin models with varying thickness is given in Fig. 6.

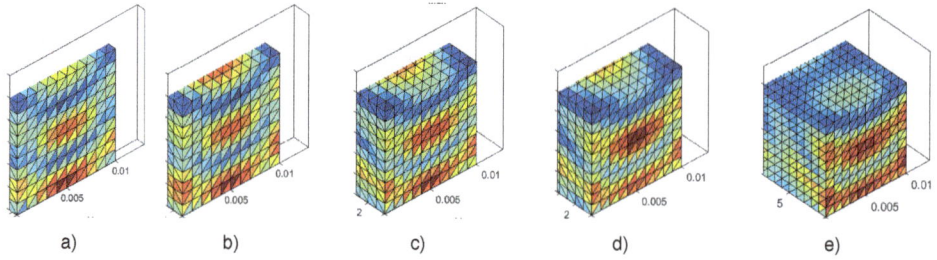

Figure 4: Distribution of equivalent electric current on the surface of planar models with varying thickness. Models at $d = 10$ mm distance to 30 GHz $\lambda/2$ dipole. (a) #1; (b) #2; (c) #4; (d) #5; and (e) #10.

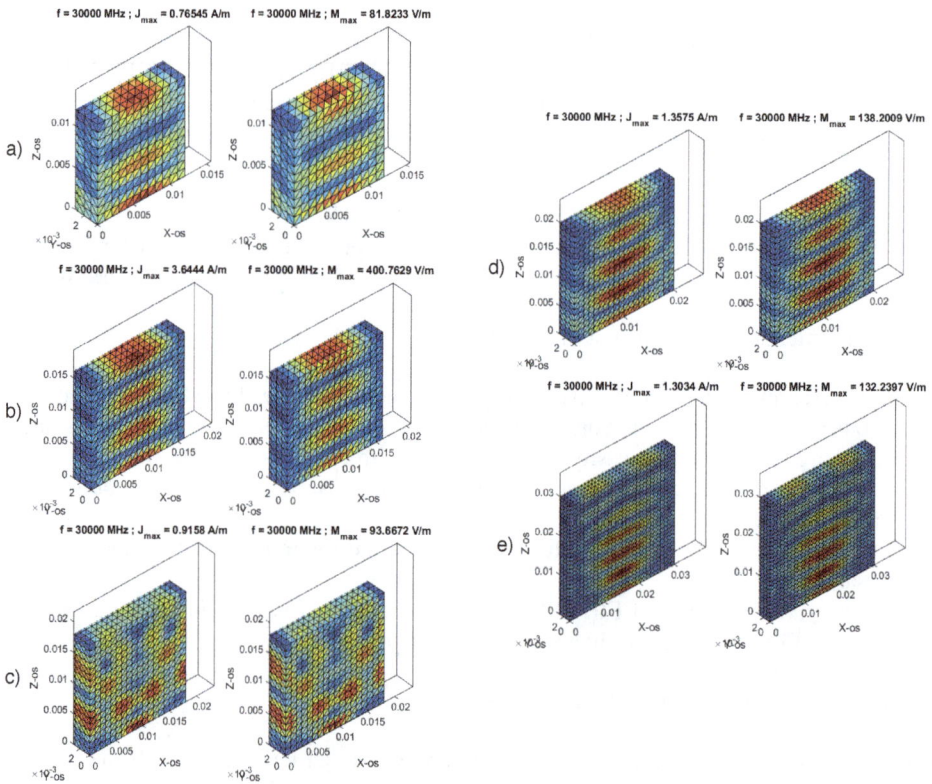

Figure 5: Equivalent electric currents and magnetic currents, respectively, on the surface of planar models with varying width. All models placed at $d = 10$ mm distance to 30 GHz $\lambda/2$ dipole. (a) #4 × 1.2; (b) #4 × 1.6; (c) #4 × 1.8; (d) #4 × 2.0; and (e) #4 × 3.0.

The authors' previous analysis of the use of planar skin models in high frequency plane wave exposure scenario has been carried out in Cvetković and Poljak [17]. The study in Cvetković and Poljak [17] showed that the reduced planar models with smaller thickness

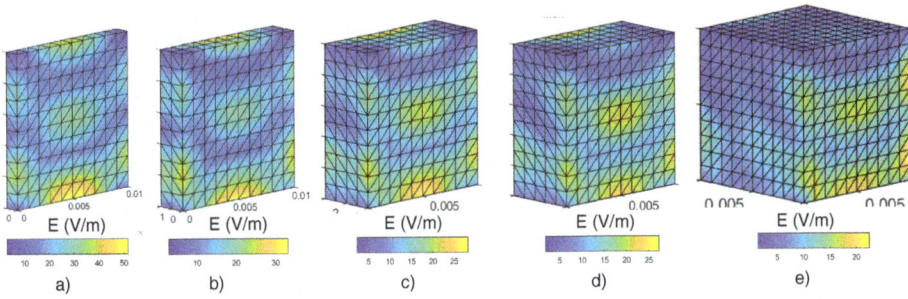

Figure 6: Induced electric field on the surface of planar models with varying thickness. Models at $d = 10$ mm distance to 30 GHz $\lambda/2$ dipole. (a) #1; (b) #2; (c) #4; (d) #5; and (e) #10.

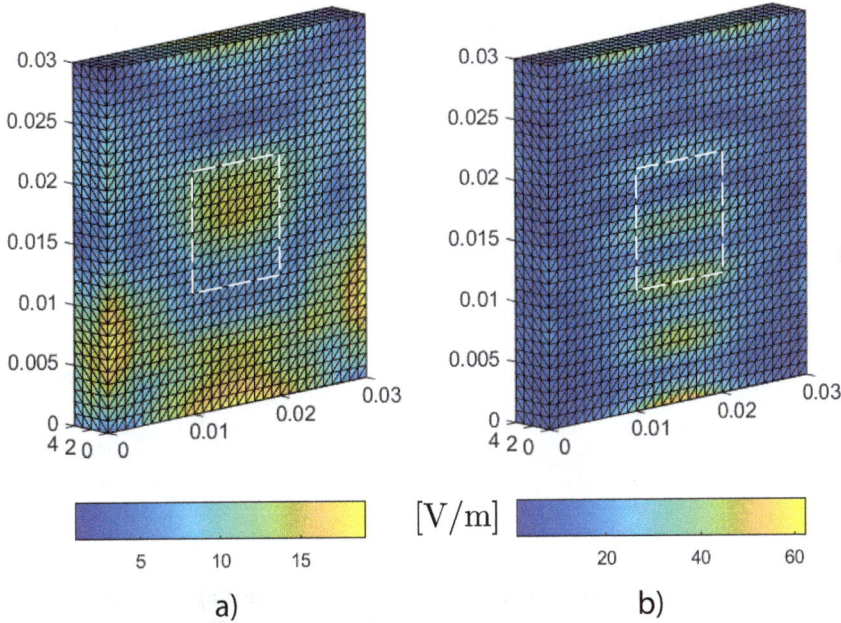

Figure 7: Electric field induced at the surface of planar model #4 × 3.0 due to $\lambda/2$ dipole at (a) 10 GHz; and (b) 30 GHz, placed at $d = 10$ mm distance. Dotted lines represent the 1 cm^2 averaging area.

could be utilized at 10 GHz, while at 30 GHz it is less obvious. Hence, a subsequent analysis of planar model depicted at Fig. 6(c) is carried out to study the effects of varying model widths.

4.4 Averaging surface area

The results for induced electric field at two frequencies of interest are shown in Fig. 7.

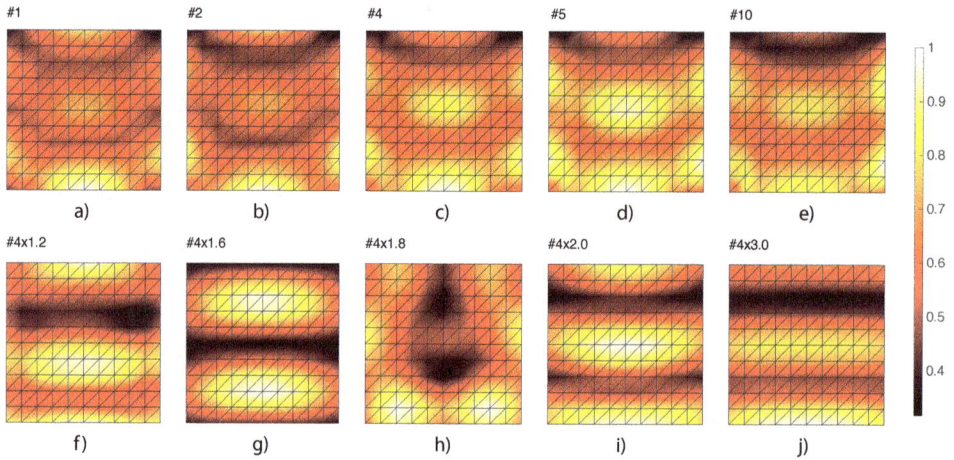

Figure 8: Normalized induced electric field at the 1 cm^2 averaging area obtained using planar models with varying thickness (a)–(e), and varying width (f)–(j). Example for 30 GHz $\lambda/2$ dipole.

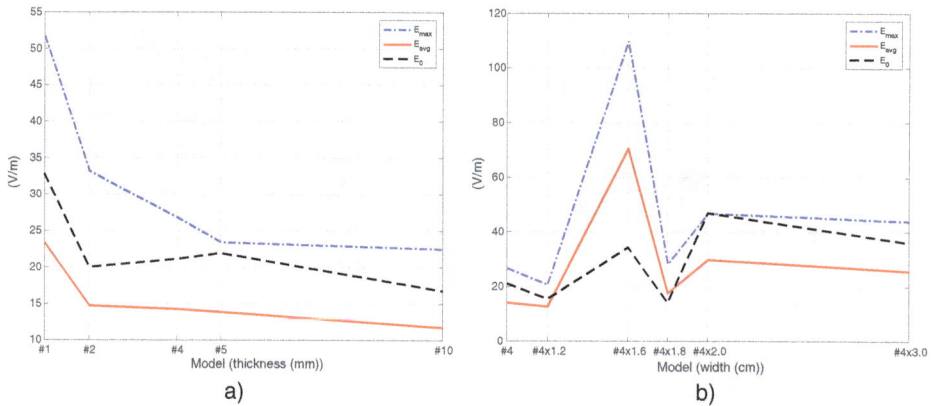

Figure 9: Dependence of several field measures from the averaging area on (a) model thickness; and (b) model width. Dotted line represents the maximum electric field (E_{max}), full line the averaged electric field (E_{avg}), and dashed line the electric field at the center point (E_0). Example for 30 GHz $\lambda/2$ dipole.

As depicted in Fig. 7, the dotted line representing the 1 cm^2 averaging area is considered when comparing models of varying thickness. The normalized distribution of the induced electric field on the 1 cm^2 surface averaging area is given in Fig. 8.

As evident from Fig. 8, generally similar field distribution is obtained using models with similar width or similar thickness, respectively. However, there are still some discrepancies, as evident e.g., for model $\#4 \times 1.8$.

More detailed comparison between used models can be found in Fig. 9, where dependence of several electric field measures from the averaging area is considered with respect to model thickness and width.

Generally, convergence of all measures are evident with the increasing model thickness (Fig. 9(a)). Compared to this, with the exception of model #4 × 1.8 results, the increase in model width results in the increase of all field measures considered on the 1 cm^2 surface averaging area.

5 CONCLUSION

The paper presented the analysis of planar skin models exposed to dipole antenna radiation in the GHz frequency range. The antenna model is based on frequency domain EFIE formulation for perfect electrically conducting body solved by means of MoM. The planar body model is based on the frequency domain EFIE formulation for homogeneous dielectric body also solved by MoM. The two models are coupled via antenna radiating electric field representing the incident field in the body model. Several numerical results were presented, including the antenna electric current, the body equivalent electric and magnetic currents, respectively, and the induced electric field at the surface of planar model. Several planar model geometries were considered. The averaging of electric field results in models with varying thickness and width showed similar field distributions. The follow-up study will be related to comparison of both SAR and APD metrics on the utilized planar models.

REFERENCES

[1] Hirata, A. et al., Assessment of human exposure to electromagnetic fields: Review and future directions. *IEEE Transactions on Electromagnetic Compatibility*, **63**(5), pp. 1619–1630, 2021.
[2] ICNIRP, Guidelines for limiting exposure to electromagnetic fields (100 kHz to 300 GHz). *Health Physics*, **118**(5), pp. 483–524, 2020.
[3] IEEE, IEEE Standard for safety levels with respect to human exposure to electric, magnetic, and electromagnetic fields, 0 Hz to 300 GHz. *IEEE Std C951-2019*, pp. 1–312, 2019.
[4] Li, K. et al., Intercomparison of calculated incident power density and temperature rise for exposure from different antennas at 10–90 GHz. *IEEE Access*, **9**, pp. 151654–151666, 2021.
[5] Li, K. et al., Calculated epithelial/absorbed power density for exposure from antennas at 10–90 GHz: Intercomparison study using a planar skin model. *IEEE Access*, **11**, pp. 7420–7435, 2023.
[6] Cvetković, M., Lojić Kapetanović, A., Poljak, D. & Dodig, H., On the applicability of numerical quadrature for double surface integrals at 5G frequencies. *Journal of Communications Software and Systems*, **18**(1), pp. 42–53, 2022.
[7] Poljak, D. & Cvetkovic, M., *Human Interaction with Electromagnetic Fields: Computational Models in Dosimetry*, Academic Press, 2019.
[8] Rao, S., Wilton, D.R. & Glisson, A., Electromagnetic scattering by surfaces of arbitrary shape. *IEEE Transactions on Antennas and Propagation*, **30**(3), pp. 409–418, 1982.
[9] Cvetković, M., Poljak, D. & Hirata, A., The electromagnetic-thermal dosimetry for the homogeneous human brain model. *Engineering Analysis with Boundary Elements*, **63**, pp. 61–73, 2016.
[10] Cvetkovic, M. & Poljak, D., Electromagnetic dosimetry based on EFIE formulation and RWG basis function. *Boundary Elements and other Mesh Reduction Methods XLI*, **122**, p. 143, 2019.
[11] Cvetković, M. & Poljak, D., Surface equivalence principle and surface integral equation (SIE) revisited for bioelectromagnetics application. *Boundary Elements and Other Mesh Reduction Methods*, p. 227, 2018.

[12] Cvetković, M., Poljak, D. & Haueisen, J., Analysis of transcranial magnetic stimulation based on the surface integral equation formulation. *IEEE Transactions on Biomedical Engineering*, **62**(6), pp. 1535–1545, 2015.

[13] Cvetković, M., Šušnjara, A. & Poljak, D., Deterministic–stochastic modeling of transcranial magnetic stimulation featuring the use of method of moments and stochastic collocation. *Engineering Analysis with Boundary Elements*, **150**, pp. 662–671, 2023.

[14] Balanis, C.A., *Antenna Theory: Analysis and Design*, John Wiley & Sons, 2016.

[15] Makarov, S.N., *Antenna and EM Modeling with MATLAB*, Wiley Inc., 2002.

[16] Poljak, D., Dorić, V. & Antonijević, S., *Computer Modeling of Wire Antennas*, Kigen doo, veljača, 2009 (in Croatian).

[17] Cvetković, M. & Poljak, D., On the use of planar skin models in high frequency dosimetry assessment. Accepted for presentation at the URSI GASS 2023, Sapporo, Japan, 19–26 Aug. 2023.

SECTION 2
DESIGN OPTIMISATION AND INVERSE PROBLEMS

FINITE ELEMENT METHOD/BOUNDARY ELEMENT METHOD-BASED MICROSTRUCTURAL TOPOLOGY OPTIMIZATION OF SUBMERGED BI-MATERIAL THIN-WALLED STRUCTURES

JIALONG ZHANG, XIAOFEI MIAO & HAIBO CHEN
CAS Key Laboratory of Mechanical Behavior and Design of Materials,
Department of Modern Mechanics, University of Science and Technology of China, China

ABSTRACT

Periodic microstructures are often found in structures in the field of vibration acoustics and topology optimization is an effective method for the design of microstructures. Microstructural topology optimization of bi-material for minimizing the responses of the exterior acoustic-structure interaction system is investigated. The structural finite element method is combined with the acoustic boundary element method to analyse the response of the acoustic-structure interaction system, and the bi-directional coupling of the acoustic-structure system is considered. The equivalent macroscopic elastic matrix of microstructures is calculated by homogenization method. Topology optimization model is schemed based on the piecewise constant level set method. Considering the high efficiency of adjoint variable method in multi-variable and high complexity optimization problems, this research adopts the adjoint variable method to analyse the sensitivity of the objective function of the coupled system. Numerical results show that the response of the coupled systems can be reduced significantly, indicating the effectiveness of the optimization algorithm.
Keywords: microstructural, topology optimization, boundary element method, finite element method, PCLS.

1 INTRODUCTION

Reducing vibration acoustic radiation of thin-walled structures is an important problem in engineering. Some structures applied in the field of vibration acoustics often have micro-structures with periodic arrangement. For such structures, the quality of the micro-structure configuration plays a decisive role in the vibration acoustic performance of the structure. Topology optimization can make full use of the potential of structure and material, and it is an effective method for the design of microstructure.

In recent years, there are a lot of researches on macroscopic topology optimization based on vibrational or acoustic criteria, but relatively few researches on material microstructure topology optimization. Yang and Du [1], [2] established a topological optimization model of material microstructure based on minimizing the radiated acoustic power of macroscopic structure surfaces. The results showed that the topological optimization of material microstructure could reduce the radiated acoustic power. They used the high frequency approximation method to get the sound pressure on the surface of the structure, avoiding solving the coupling equation. In addition, in the case of air, the effect of the sound field on the structure can be ignored, so they used the weak coupling condition. Considering the strong coupling between structure and sound field, Chen et al. [3], [4] established the robust micro-structure topology optimization model of the internal sound field coupling system by using finite element method (FEM). However, for the external sound field, it is difficult to apply the boundary conditions and the discretization of the field is too large by using FEM. For the exterior acoustic-structure interaction system, the FEM/boundary element method (BEM) coupling algorithm is a better solution, Zhao et al. [5] proposed a bi-material topology

WIT Transactions on Engineering Sciences, Vol 135, © 2023 WIT Press
www.witpress.com, ISSN 1743-3533 (on-line)
doi:10.2495/BE460031

optimization approach based this method. Following this line, a microstructural topology optimization method based on FEM/BEM is proposed in this research.

2 STRUCTURAL-ACOUSTIC ANALYSIS

The macroscopic equivalent elastic constant matrix D^H of microstructure can be obtained by the homogenization method [6], [7].

$$D^H = \frac{1}{|Y|} \sum_{e=1}^{n_e} \int_{Y_e} D^{MI}(I - bu_e)dY, \tag{1}$$

where $|Y|$ denotes the volume of a microcell, n_e denotes the total number of cell units, D^{MI} denotes the element elastic matrix of a cell, I denotes the identity matrix, b denotes the strain matrix of a cell, u_e denotes the element displacement matrix of a cell.

The equivalent material elastic matrix D^H can be incorporated into the FEM part of the FEM/BEM coupling method. And the derivative part can be calculated by substituting the following formula into the FEM part.

$$\frac{\delta D^H}{\delta x} = \frac{1}{|Y|} (\int_{Y_e} (I - bu_e)^T \frac{\delta D^H}{\delta x}(I - bu_e)dY). \tag{2}$$

Considering the impact of sound field on the structure, the finite element discrete equation of the structure is

$$(K - \omega^2 M)u = Au = f + C_{sf}p, \tag{3}$$

where K and M denote respectively the stiffness and the mass matrices of the structure, u denotes the nodal displacement vector, f denotes the nodal force vector and $C_{sf} = \int N_s^T n N_f \, d\Gamma$ denotes the coupling matrix, N_f and N_s denote the shape functions for the fluid and structural domains, and n is the unit normal vector on the boundary Γ. p denotes the sound pressure vector. The acoustic domain which is based on the BEM formulation can be written as

$$Hp = Gq + p_{in}, \tag{4}$$

where H and G are respectively the BEM coefficient matrices, q and p_{in} denote the flux and incident wave vectors. The governing equations, eqns (3) and (4), are linked via the continuity condition:

$$v = -i\omega S^{-1} C_{fs} u, \tag{5}$$

where v is the normal vibration velocity of the acoustic medium at the boundary Γ, $S = \int N_f^T N_f \, d\Gamma$, and $C_{fs} = C_{sf}^T$. According to $q = i\omega\rho v$, eqns (3) and (4) can be written as the following formulation

$$(H - GWC_{sf})p = GWf_s + p_{in}, \tag{6}$$

where $W = \omega^2 \rho S^{-1} C_{fs} A^{-1}$. The iterative solver GMRES and PARDISO package in Intel Math Kernel Library (MKL) are used to solve the above equation. After solving the equation, the acoustic and structural response is obtained. After discretization, the sound power P can be expressed as follows

$$P = \frac{1}{2}\Re(p^T S v^*), \tag{7}$$

where $\Re$ denotes the real part of complex value, and $()^*$ denotes the conjugate transpose.

3 MINIMIZATION OF VIBRO-ACOUSTICS RESPONSE USING TOPOLOGY OPTIMIZATION

The vibro-acoustics response of coupling system can be described by structural displacement u and sound field pressure p or a function of u and p. The sound power level is taken as the objective function. The problem can be formulated as follows:

$$\min \Pi = 10\log\frac{P}{P_0}$$
$$\text{s.t.} \sum_{e=1}^{N} \rho_e V_e - f_v \sum_{e=1}^{N} V_e \leq 0,$$
$$\rho_e \in \{0,1\}$$

(8)

where P_0 is the reference sound power, ρ_e is the design variable, 0 and 1 represent the regions of material 1 and material 2, f_v denotes the volume fraction constraint.

Details of calculating sensitivity using adjoint variable method can be referred to [5]. The piecewise constant level set method is employed to solve the optimization problem [8]. When the relative change of the objective function value is less than a specified value, the iteration terminates.

4 NUMERICAL EXAMPLES

A numerical example is presented to illustrate the validity of the optimization method in this section. Considering a submersed hexahedral thin-walled box as shown in Fig. 1. The dimensions of the cube are 1 m × 1 m × 1 m with the thickness 0.01m. The top surface of the cube discretized by Kirchhoff plate elements and the other surfaces are rigid. The steel is chosen as the Material 1 (E_1=210 GPa, $\rho_1 = 7800$kg/m^3, $v_2 = 0.3$), and Material 2 has the properties $E_2 = 0.1E_1$, $\rho_2 = 0.1\rho_1$, and $v_2 = v_1$. The harmonic excitation is set to be F = 1,000 N, $f_p = 60$ Hz, and the volume ratio of material 1 to material 2 is 1:1.

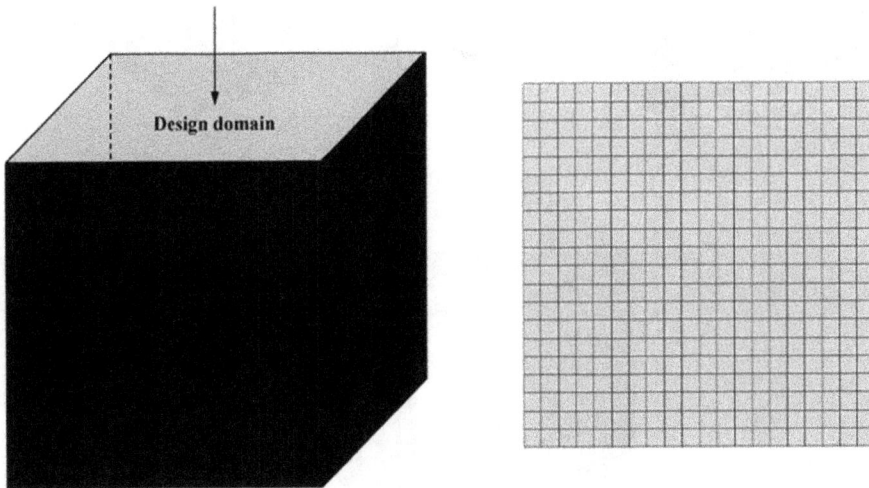

Figure 1: The hexahedral structure, model and mesh.

Fig. 2 shows the optimized microstructural unit cells, and Fig. 3 shows the sound pressure level (SPL) of the structure surface. As can be seen from Fig. 3 that the SPL on the surface of the structure is reduced after optimization, and even lower than that when all material 1 with higher stiffness are used, which indicates that the optimization is effective.

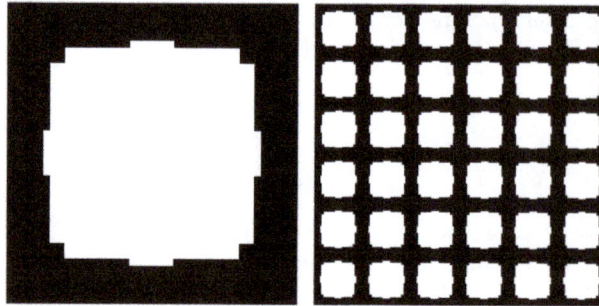

Figure 2: The optimized microstructural unit cell and the corresponding 6 × 6 arrays.

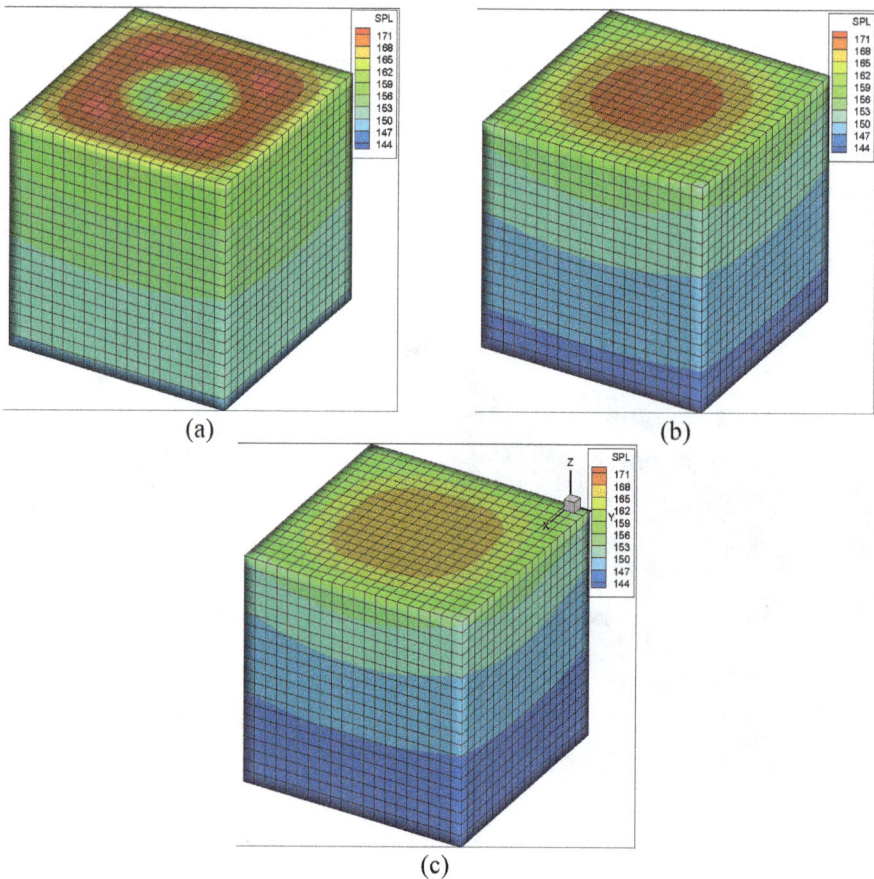

(a)

(b)

(c)

Figure 3:   The sound pressure level of the structural surface. (a) Before optimization; (b) Full of material 1; and (c) After optimization.

5 CONCLUSIONS

In this paper, the microstructural topology optimization technique for bi-material structure has been presented to find the optimum microstructural distribution which reduces the responses of the exterior acoustic-structure interaction systems. The effect of sound field on the structure is considered in optimization. Numerical examples show that this method can reduce the response of acoustic-structure interaction systems.

ACKNOWLEDGEMENT

This work was financially supported by the National Natural Science Foundation of China (NSFC) under Grant No. 12172350.

REFERENCES

[1] Yang, R. & Du, J., Microstructural topology optimization with respect to sound power radiation. *Structural and Multidisciplinary Optimization*, **47**, pp. 191–206, 2013.

[2] Du, J. & Yang, R., Vibro-acoustic design of plate using bi-material microstructural topology optimization. *Journal of Mechanical Science and Technology*, **29**, pp. 1413–1419, 2015.

[3] Chen, N. et al., Microstructural topology optimization of structural-acoustic coupled systems for minimizing sound pressure level. *Structural and Multidisciplinary Optimization*, 56, pp. 1259–1270, 2017.

[4] Chen, N. et al., Microstructural topology optimization for minimizing the sound pressure level of structural-acoustic systems with multi-scale bounded hybrid uncertain parameters. *Mechanical Systems and Signal Processing*, **134**, pp. 106336, 2019.

[5] Zhao, W. et al., Minimization of sound radiation in fully coupled structural-acoustic systems using FEM-BEM based topology optimization. *Structural and Multidisciplinary Optimization*, **58**, pp. 115–128, 2018.

[6] Bendsøe, M.P. & Kikuchi, N., Generating optimal topologies in structural design using a homogenization method. *Computer Methods in Applied Mechanics and Engineering*, **71**(2), pp. 197–224, 1988.

[7] Bendsøe, M.P., Díaz, A. & Kikuchi, N., Topology and generalized layout optimization of elastic structures. *Topology Design of Structures*, **227**, pp. 159–206, 1993.

[8] Zhang, Z. & Chen, W., An approach for topology optimization of damping layer under harmonic excitations based on piecewise constant level set method. *Journal of Computational Physics*, **390**, pp. 470–489, 2019.

TREATMENT OF TOPOLOGY OPTIMIZATION OF A TWO-DIMENSIONAL FIELD GOVERNED BY LAPLACE'S EQUATION UNDER NONLINEAR BOUNDARY CONDITION

SHINSEI SATO, YI CUI, TORU TAKAHASHI & TOSHIRO MATSUMOTO
Department of Mechanical Science and Engineering, Nagoya University, Japan

ABSTRACT

This paper presents a treatment of the topology optimization problem for two-dimensional fields governed by Laplace's equation. The study considers various boundary conditions, including Dirichlet, Neumann, Robin, and nonlinear radiation boundary conditions. Additionally, the topological derivative for a general objective functional comprising solely of boundary quantities is derived, with a special focus on the case of a radiation boundary condition in a black body. The accuracy of the derived adjoint problem and topological derivative is validated through several boundary element method calculations.

Keywords: *topology optimization, nonlinear boundary condition, Laplace's equation, adjoint problem, topological derivative.*

1 INTRODUCTION

The level-set based methods [1], [2] have been proposed as one of the robust topology optimization techniques and have been widely and successfully applied in various engineering applications. The primary successful application of topology optimization has been in the minimization of mean compliance for linear elastostatic bodies under linear boundary conditions, such as fixed and traction boundaries.

However, there are some structural optimization problems involving linear governing equations but under nonlinear boundary conditions, such as those encountered in galvanic cathodic protection [3], [4] and thermal radiation problems [5]–[8].

The level-set based methods utilize a continuous scalar function called the 'level set function', defined for points within a fixed design domain, to determine the region in which the material exists. This function assigns a positive value to points inside the material region, a zero value to boundary points, and a negative value to points outside the material region. Consequently, the actual material distribution can be extracted from the iso-surface of the level set function corresponding to the value zero.

The distribution of the level-set function is updated by solving an evolution equation [2], which is a type of reaction-diffusion equation. This evolution equation is accompanied by a source term corresponding to the 'topological derivative'. The topological derivative represents the sensitivity of the objective functional when an infinitesimal region is removed from the material region. The calculation of the topological derivative requires solving both the original problem and an adjoint problem.

In this study, we consider a structural optimization problem for boundary value problems governed by Laplace's equation with both linear and nonlinear boundary conditions. The variation of the objective functional, defined by a boundary integral of the potential and flux, is derived by appropriately defining an adjoint problem. Some numerical demonstrations are presented for both general and radiation-type nonlinear boundary conditions.

WIT Transactions on Engineering Sciences, Vol 135, © 2023 WIT Press
www.witpress.com, ISSN 1743-3533 (on-line)
doi:10.2495/BE460041

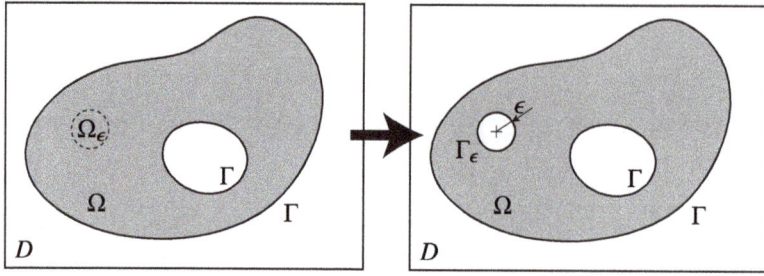

Figure 1: An infinitesimal circular region Ω_ϵ is removed from the material region Ω.

2 STATEMENT OF TOPOLOGY OPTIMIZATION

2.1 Level set method

We assume that a scalar function $\phi(\boldsymbol{x})$ is defined in $D \subset \mathbb{R}^2$. The material region Ω and its boundary Γ are assumed to be included in D, as follows:

$$\Omega = \{ \boldsymbol{x} \mid 0 < \phi(\boldsymbol{x}) \leq 1 \}, \tag{1}$$

$$\Gamma = \{ \boldsymbol{x} \mid \phi(\boldsymbol{x}) = 0 \}, \tag{2}$$

$$D \setminus \overline{\Omega} = \{ \boldsymbol{x} \mid -1 \leq \phi(\boldsymbol{x}) < 0 \}. \tag{3}$$

From this definition of $\phi(\boldsymbol{x})$, the material region can be described based on the distribution of $\phi(\boldsymbol{x})$. By changing the distribution of $\phi(\boldsymbol{x})$, the material distribution also changes accordingly. The level-set method of [2] assumes that the distribution of $\phi(\boldsymbol{x})$ is the solution of the following initial and boundary value problem:

$$\frac{\partial \phi(\boldsymbol{x},t)}{\partial t} = -K J_T(\boldsymbol{x},t) + \tau \nabla^2 \phi(\boldsymbol{x},t) \quad \boldsymbol{x} \in D, \ t > 0, \tag{4}$$

$$\phi(\boldsymbol{x},0) = \phi_0 \quad \boldsymbol{x} \in D, \tag{5}$$

$$\phi(\boldsymbol{x},t) = c \quad \boldsymbol{x} \in \partial D_+, \ t > 0, \tag{6}$$

$$\phi(\boldsymbol{x},t) = -c \quad \boldsymbol{x} \in \partial D_-, \ t > 0, \tag{7}$$

where t is a fictitious time, K is a positive constant, τ is also a positive constant regularization parameter for blurring the distribution of ϕ, and ϕ_0 is the initial distribution of ϕ corresponding to the initial geometry of the material region Ω at $t = 0$. ∂D_+ and ∂D_- are parts of the boundary ∂D of the domain D, where $\partial D_+ \cup \partial D_- = \partial D$. The Dirichlet condition (6) with a positive constant $c \in (0, 1]$ restricts that the material always exists on ∂D_+, while (7) with a negative constant $-c \in [-1, 0)$ allows the material removal in D in the neighborhood of ∂D_-. J_T denotes the topological derivative, defined by

$$\delta J(\boldsymbol{x}) = J_T(\boldsymbol{x}) b(\epsilon) + o(b(\epsilon)), \tag{8}$$

where $\delta J(\boldsymbol{x})$ is the variation of J when an infinitesimal circular region of radius ϵ centered at $\boldsymbol{x}$ is removed as shown in Fig. 1, and $b(\epsilon)$ is a constant that vanishes as $\epsilon \to 0$. After $\delta J(\boldsymbol{x})$ is expanded analytically in the above form, the topological derivative is obtained by

$$J_T(\boldsymbol{x}) = \lim_{\epsilon \to 0} \frac{\delta J(\boldsymbol{x})}{b(\epsilon)}. \tag{9}$$

WIT Transactions on Engineering Sciences, Vol 135, © 2023 WIT Press
www.witpress.com, ISSN 1743-3533 (on-line)

2.2 Boundary value problem with a nonlinear boundary condition

Let us consider a boundary value problem with a general nonlinear boundary condition as follows:

$$\nabla \cdot (-k\nabla u) = 0 \quad \text{in} \quad \Omega, \tag{10}$$

$$u = \bar{u} \quad \text{on} \quad \Gamma_u, \tag{11}$$

$$q := -k\nabla u \cdot \boldsymbol{n} = \bar{q} \quad \text{on} \quad \Gamma_q, \tag{12}$$

$$q := -k\nabla u \cdot \boldsymbol{n} = \nu(u) \quad \text{on} \quad \Gamma_n, \tag{13}$$

where u is potential, k is a positive constant, both $\bar{u}$ and $\bar{q}$ are known functions, $\boldsymbol{n}$ is the unit outward normal vector to the boundary, and $\nu(u)$ is a nonlinear function of u. For thermal radiation problems of a black body, $\nu(u)$ can be given as

$$\nu(u) = \sigma u^4, \tag{14}$$

where σ is the Stefan–Boltzmann constant. In eqn (13), the interactions of the radiations are not taken into account for simplicity.

2.3 Objective functional and topological derivative

We consider the following augmented objective functional to minimize:

$$J = \int_{\Gamma} f(u, q) \, d\Gamma + \int_{\Omega} \mu \nabla \cdot (-k\nabla u) \, d\Omega$$

$$= \int_{\Gamma} f(u, q) \, d\Gamma + \int_{\Gamma} \mu q \, d\Gamma - \int_{\Omega} \nabla\mu \cdot (-k\nabla u) \, d\Omega, \tag{15}$$

where μ is an adjoint variable.

By removing an infinitesimal circular region Ω_ϵ from Ω, the objective functional may change from J to $J + \delta J$, and after some lengthy operations, we obtain

$$J + \delta J = \int_{\Gamma_u \cup \Gamma_q \cup \Gamma_n} (f + \delta f) \, d\Gamma + \int_{\Gamma_\epsilon} (f + \delta f) \, d\Gamma$$

$$+ \int_{\Gamma_u \cup \Gamma_q \cup \Gamma_n} \mu(q + \delta q) \, d\Gamma + \int_{\Gamma_\epsilon} \mu(q + \delta q) \, d\Gamma$$

$$- \int_{\Omega \setminus \Omega_\epsilon} (-k\nabla\mu) \cdot \nabla(u + \delta u) \, d\Omega. \tag{16}$$

Subtracting both sides of eqn (15) from (16) results in

$$\delta J = \int_{\Gamma_u} \left(\mu + \frac{\partial f}{\partial q} \right) \delta q \, d\Gamma - \int_{\Gamma_q} \left(\eta - \frac{\partial f}{\partial u} \right) \delta u \, d\Gamma$$

$$- \int_{\Gamma_n} \left[\left(\eta - \frac{\partial f}{\partial u} \right) \delta u - \left(\mu + \frac{\partial f}{\partial q} \right) \delta q \right] d\Gamma + \int_{\Gamma_\epsilon} (f + \delta f) \, d\Gamma$$

$$+ \int_{\Gamma_\epsilon} \mu(q + \delta q) \, d\Gamma - \int_{\Gamma_\epsilon} \eta \, \delta u \, d\Gamma$$

$$+ \int_{\Omega \setminus \Omega_\epsilon} \nabla \cdot (-k\nabla\mu) \, \delta u \, d\Omega + \int_{\Omega_\epsilon} (-k\nabla\mu) \cdot \nabla u \, d\Omega, \tag{17}$$

where $\eta = -k\nabla\mu \cdot \boldsymbol{n}$.

On Γ_n, we find that $\delta q = \frac{\partial \nu(u)}{\partial u} \delta u$, thus, eqn (18) can be rearranged as

$$\delta J = \int_{\Gamma_u} \left(\mu + \frac{\partial f}{\partial q} \right) \delta q \, d\Gamma - \int_{\Gamma_q} \left(\eta - \frac{\partial f}{\partial u} \right) \delta u \, d\Gamma$$

$$- \int_{\Gamma_n} \left[\eta - \frac{\partial f}{\partial u} - \left(\mu + \frac{\partial f}{\partial q} \right) \frac{\partial \nu}{\partial u} \right] \delta u \, d\Gamma + \int_{\Gamma_\epsilon} (f + \delta f) \, d\Gamma$$

$$+ \int_{\Gamma_\epsilon} \mu(q + \delta q) \, d\Gamma - \int_{\Gamma_\epsilon} \eta \, \delta u \, d\Gamma$$

$$+ \int_{\Omega \setminus \Omega_\epsilon} \nabla \cdot (-k \nabla \mu) \, \delta u \, d\Omega + \int_{\Omega_\epsilon} (-k \nabla \mu) \cdot \nabla u \, d\Omega. \tag{18}$$

Now, we employ μ, which is the solution of the following adjoint boundary value problem.

$$\nabla \cdot (-k \nabla \mu) = 0 \quad \text{in} \quad \Omega, \tag{19}$$

$$\mu = -\frac{\partial f}{\partial q} \quad \text{on} \quad \Gamma_u, \tag{20}$$

$$\eta := -k \nabla \mu \cdot \boldsymbol{n} = \frac{\partial f}{\partial u} \quad \text{on} \quad \Gamma_q, \tag{21}$$

$$\eta := -k \nabla \mu \cdot \boldsymbol{n} = \frac{\partial \nu}{\partial u} \mu + \frac{\partial f}{\partial u} + \frac{\partial \nu}{\partial u} \frac{\partial f}{\partial q} \quad \text{on} \quad \Gamma_n. \tag{22}$$

In the above boundary value problem, we find that the boundary condition for Γ_n is no longer a nonlinear boundary condition but a type of Robin boundary condition. By using the solution of the above boundary value problem, the variation of the objective functional can be obtained as

$$\delta J = \int_{\Gamma_\epsilon} (f + \delta f) \, d\Gamma + \int_{\Gamma_\epsilon} \mu(q + \delta q) \, d\Gamma$$

$$- \int_{\Gamma_\epsilon} \eta \, \delta u \, d\Gamma + \int_{\Omega_\epsilon} (-k \nabla \mu) \cdot \nabla u \, d\Omega, \tag{23}$$

where the boundary condition for Γ_ϵ is also assumed to be a nonlinear one, i.e.,

$$q + \delta q = \nu(u + \delta u) \approx \nu(u) + \frac{\partial \nu}{\partial u} \delta u. \tag{24}$$

We consider the asymptotic behaviors of δu and δq, and the Taylor series expansions of u and q about the center of Ω_ϵ. After some lengthy operations, we find

$$\delta u \Big|_{\Gamma_\epsilon} \approx \epsilon \left(\frac{\partial u}{\partial x}(\boldsymbol{x}_0) \cos \theta + \frac{\partial u}{\partial y}(\boldsymbol{x}_0) \sin \theta \right), \tag{25}$$

and

$$q + \delta q \Big|_{\Gamma_\epsilon} \approx \nu(\boldsymbol{x}_0) + 2\epsilon \frac{\partial \nu}{\partial u}(\boldsymbol{x}_0) \left(\frac{\partial u}{\partial x}(\boldsymbol{x}_0) \cos \theta + \frac{\partial u}{\partial y}(\boldsymbol{x}_0) \sin \theta \right), \tag{26}$$

where $\boldsymbol{x}_0$ is the center of the circular region Ω_ϵ.

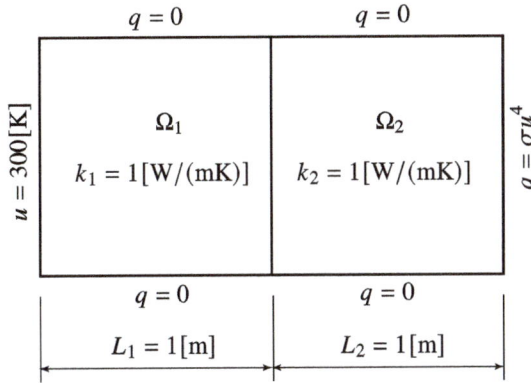

Figure 2: A domain consisting of two squares Ω_1 and Ω_2 subjected to a radiation boundary condition.

After evaluating the integrals for Γ_ϵ and Ω_ϵ, we obtain the variation of J as follows:

$$\delta J(\boldsymbol{x}_0) = \int_{\Gamma_\epsilon} (f + \delta f)\, d\Gamma + 2\pi\epsilon\mu(\boldsymbol{x}_0)\nu(\boldsymbol{x}_0) - 2\pi\epsilon^2 k\nabla\mu(\boldsymbol{x}_0) \cdot \nabla u(\boldsymbol{x}_0) + o(\epsilon^2). \quad (27)$$

Therefore, the topological derivative at a point $\boldsymbol{x}$ in Ω is obtained as

$$J_T(x) = \lim_{\epsilon \to 0} \frac{\delta J}{2\pi\epsilon} = \lim_{\epsilon \to 0} \int_{\Gamma_\epsilon} (f + \delta f)\, d\Gamma + \mu(\boldsymbol{x})\nu(\boldsymbol{x}). \quad (28)$$

The actual expression of the first term integral of J_T depends of the definition of the boundary objective functional f defined on the newly generated boundaries, but it should be chosen so that this limit can exist.

3 NUMERICAL EXAMPLE

We consider a steady-state heat conduction problem for a domain consisting of two squares with a side length of 1 m as shown in Fig. 2. The thermal conductivities of both domains are assumed to be the same, with $k_1 = k_2 = 1$ (W/(mK)). On the left-side boundary, the temperature $u = 300$ (K) is given, and on the right-side, the radiation condition $q = \sigma u^4$ is applied. The top and bottom boundaries are assumed to be adiabatic. On the interface boundary between Ω_1 and Ω_2, the temperature is continuous, and the heat fluxes are balanced. We discretized each edge with 10 constant boundary elements and calculated the temperatures and the adjoint temperatures on the boundaries and at twenty uniformly arranged internal points along the center line in the x direction. In Fig. 3, we show the temperature distributions calculated by using the boundary element method at the internal points and on the boundaries along the center lines in the x-direction, comparing them with the exact solutions. Both the BEM and exact solutions agree very well. The boundary element method required 48 iterations based on the fixed-point iteration approach until the temperature on the radiation boundary converged.

WIT Transactions on Engineering Sciences, Vol 135, © 2023 WIT Press
www.witpress.com, ISSN 1743-3533 (on-line)

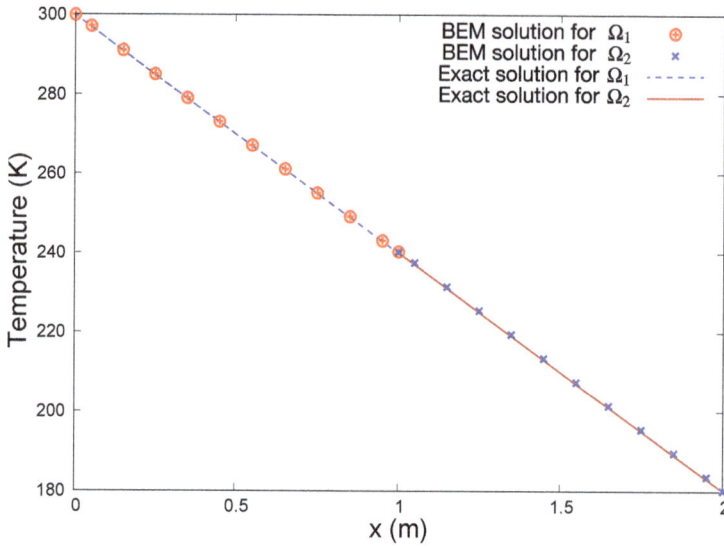

Figure 3: An infinitesimal circular region Ω_ϵ is removed from the material region Ω.

Next, we calculated the topological derivatives by using the expression of eqn (28). The boundary objective functional is defined as $f(u, q) = (u - \tilde{u})^2$, where $\tilde{u} = 300$ (K). Then, the adjoint boundary value problem becomes as follows:

$$\nabla \cdot (-k\nabla\mu) = 0 \quad \text{in} \quad \Omega, \tag{29}$$

$$\mu = 0 \quad \text{on} \quad \Gamma_u, \tag{30}$$

$$\eta = 2(u - \tilde{u}) \quad \text{on} \quad \Gamma_q, \tag{31}$$

$$\eta = 3\sigma u^3\mu + 2(u - \tilde{u}) \quad \text{on} \quad \Gamma_n. \tag{32}$$

No objective functional is assumed on Γ_ϵ for simplicity; hence, the topological derivative becomes as follows:

$$J_T(x) = \mu(\boldsymbol{x})\nu(\boldsymbol{x}) = 3\sigma u(\boldsymbol{x})^3\mu(\boldsymbol{x}). \tag{33}$$

In order to validate the topological derivative, we compared the values calculated by eqn (33) at 9×9 points uniformly arranged in each square region with those obtained by finite difference approximations of the objective functional for the original domain and the domain from which a small circular region is removed, i.e.,

$$J_T(\boldsymbol{x}) \approx \frac{J_\Omega - J_{\Omega \setminus \Omega_\epsilon}(\boldsymbol{x})}{2\pi\epsilon}, \tag{34}$$

where the radius of Ω_ϵ was set as $\epsilon = 0.01$ (m).

The boundary of the circular cavity Γ_ϵ was divided into 10 constant boundary elements, and the objective functional $\int_\Gamma (u - \tilde{u})^2 \, d\Gamma$ was calculated using the temperature values obtained by the BEM. In Fig. 4, we show the topological derivative values calculated using the analytical expression of eqn (33) and the approximate topological derivative values calculated using eqn (34). Both results show very good agreement, confirming the correctness of the representation of the newly derived topological derivative.

WIT Transactions on Engineering Sciences, Vol 135, © 2023 WIT Press
www.witpress.com, ISSN 1743-3533 (on-line)

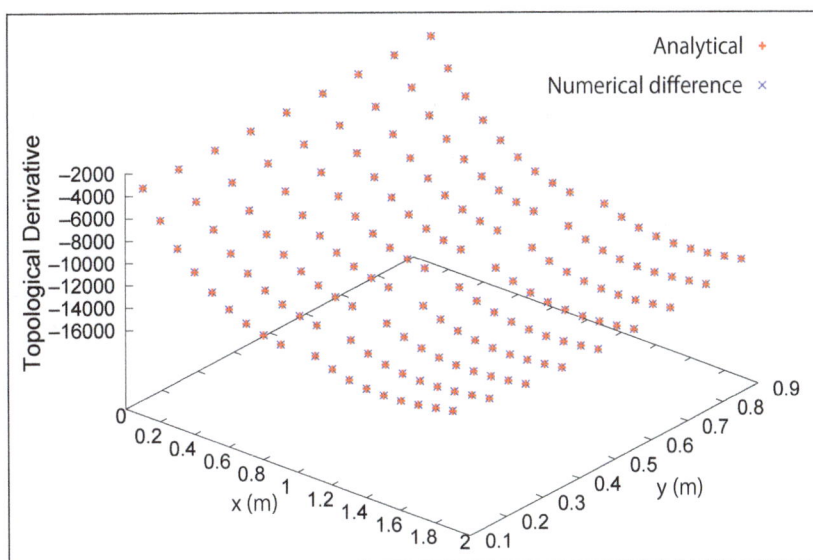

Figure 4: An infinitesimal circular region Ω_ϵ is removed from the material region Ω.

4 CONCLUSION

We considered topology optimization problems for a field governed by boundary value problems of Laplace's equation with nonlinear boundary conditions and derived the analytical expression of the topological derivative. Several numerical examples were demonstrated to validate the boundary element method and the topological derivative expression, particularly when a thermal radiation boundary condition was specified on some part of the boundary.

The topological derivative plays a crucial role in level-set based topology optimization, serving as the source term in the reaction-diffusion equation of the level set function. An extension of the current topological derivative expression is necessary to include the radiation condition that accounts for the thermal radiation from other boundaries.

ACKNOWLEDGEMENT

This work was supported by JSPS KAKENHI Grant Number JP19H00740.

REFERENCES

[1] Wang, S.Y., Lim, K.M., Khoo, B.C. & Wang, M.Y., An extended level set method for shape and topology optimization. *Journal of Computational Physics*, **221**, pp. 395–421, 2007.

[2] Yamada, T., Izui, K., Nishiwaki, S. & Takezawa, A., A topology optimization method based on the level set method incorporating a fictitious interface energy. *Computer Methods in Applied Mechanics and Engineering*, **199**, pp. 2876–2891, 2010.

[3] Aoki, S. & Kishimoto, K., Application of BEM to galvanic corrosion and cathodic protection. *Electrical Engineering Applications, Topics in Boundary Element Research*, ed. C.A. Brebbia, Springer: Berlin, Heidelberg, vol. 7, 1990.

[4] Zamani, N.G., Chuang, J.M. & Porter, J.F., BEM simulation of cathodic protection systems employed in infinite electrolytes. *International Journal for Numerical Methods in Engineering*, **24**(3), pp. 477–662, 1987.

[5] Skerget, P. & Alujevic, A., Boundary element method in nonlinear transient heat transfer of reactor solids with convection and radiation on surfaces. *Nuclear Engineering and Design*, **76**, pp. 47–54, 1983.

[6] Keavey, M.A., An isoparametric boundary solution for thermal radiation. *Communications in Applied Numerical Methods*, **4**, pp. 639–646, 1988.

[7] Blobner, J., Białecki, R. & Kuhn, G., Boundary-element solution of coupled heat conduction-radiation problems in the presence of shadow zones. *Numerical Heat Transfer, Part B: Fundamentals, An International Journal of Computation and Methodology*, **39**, pp. 451–478, 2001.

[8] Białecki, R.A., Burczyński, T., Długosz, A., Kuś, W. & Ostrowski, Z., Evolutionary shape optimization of thermoelastic bodies exchanging heat by convection and radiation. *Computer Methods in Applied Mechanics and Engineering*, **194**, pp. 1839–1859, 2005.

SECTION 3
ADVANCED FORMULATIONS

QUADRATURE RULE FOR SOLVING THE HELMHOLTZ EQUATION IN HYPERSINGULAR BEM FORMULATION

ANTONIO ROMERO[1], ROCIO VELÁZQUEZ-MATA[1], JOSE DOMÍNGUEZ[1],
ANTONIO TADEU[2,3] & PEDRO GALVÍN[1,4]
[1]Escuela Técnica Superior de Ingeniería, Universidad de Sevilla, Spain
[2]University of Coimbra, CERIS, Department of Civil Engineering, Portugal
[3]Itecons, Institute of Research and Technological Development in Construction, Energy,
Environment and Sustainability, Portugal
[4]ENGREEN, Laboratory of Engineering for Energy and Environmental Sustainability,
Universidad de Sevilla, Spain

ABSTRACT

Velázquez-Mata et al. [1] recently presented a quadrature rule to accurately evaluate singular and weakly singular integrals in the sense of the Cauchy Principal Value by an exclusively numerical procedure. The procedure was verified by solving engineering problems using the boundary element method with fundamental solutions that have singularities of type $\log(r)$ and $1/r$. However, that quadrature does not handle the evaluation of the Hadamard Finite Part of hypersingular integrals. These types of singularity appear in several fundamental solutions and, also, when the hypersingular boundary element formulation is applied to the Green functions previously analysed by the authors. In this paper, the quadrature rule presented in Velázquez-Mata et al. [1] is extended to accurately compute integrals with singularities of the type $1/r^2$. The quadrature weights are derived from a system of equations defined from the finite part of known integrals called generalised moments, which include the element shape functions. This novelty is included in the hypersingular formulation of the boundary element method to solve the Helmholtz equation, taking advantage of this methodology to consider null-thickness boundaries using the Dual BEM.

Keywords: hypersingular formulation, dual BEM, boundary integral equation, hypersingular kernels, singular kernels, Bézier curve.

1 INTRODUCTION

The formulation of the classical boundary element method (BEM) results in boundary integral equations (BIE) for a point $\mathbf{x}^*$ located at the arbitrary boundary Γ as follows [2]:

$$c(\mathbf{x}^*)u(\mathbf{x}^*) = \int_{\Gamma} \left(t(\mathbf{x})\mathcal{G}(\mathbf{x}, \mathbf{x}^*) - u(\mathbf{x})\mathcal{H}(\mathbf{x}, \mathbf{x}^*) \right) d\Gamma(\mathbf{x}), \tag{1}$$

where $u(\mathbf{x})$ and $t(\mathbf{x})$ are the field variables, $\mathcal{G}(\mathbf{x}, \mathbf{x}^*)$ and $\mathcal{H}(\mathbf{x}, \mathbf{x}^*)$ are the fundamental solution at the point $\mathbf{x}$ due to a point source located at $\mathbf{x}^*$, and the integral-free term $c(\mathbf{x}^*)$ depends only on the boundary geometry at the collocation point $\mathbf{x}^*$.

The hypersingular formulation is much better suited for some applications, such as when minimal thickness should be modelled. The hypersingular boundary element formulation or the traction boundary element method (TBEM) is obtained by taking the boundary integral gradient in eqn (1):

$$c(\mathbf{x}^*)t(\mathbf{x}^*) + a(\mathbf{x}^*)u(\mathbf{x}^*) = \int_{\Gamma} (t(\mathbf{x})\tilde{\mathcal{G}}(\mathbf{x}, \mathbf{x}^*) - u(\mathbf{x})\tilde{\mathcal{H}}(\mathbf{x}, \mathbf{x}^*)) d\Gamma(\mathbf{x}), \tag{2}$$

where the coefficients $a(\mathbf{x}^*)$ are zeros in the cases indicated by Guiggiani [3] that include the ones considered in this paper, and $\tilde{\mathcal{G}}(\mathbf{x}, \mathbf{x}^*)$ and $\tilde{\mathcal{H}}(\mathbf{x} \text{ and } \mathbf{x}^*)$ are obtained from the derivatives of the fundamental solution.

WIT Transactions on Engineering Sciences, Vol 135, © 2023 WIT Press
www.witpress.com, ISSN 1743-3533 (on-line)
doi:10.2495/BE460051

The integrals in eqns (1) and (2) can be regular, near-singular, weakly singular, singular, or hypersingular integrals and should be understood in the sense of the Cauchy Principal Value (CPV) or in the Hadamard Finite Part (FP). It should be indicated that the order of the singularity increases in the hypersingular formulation as the gradient is taken. This work extends the quadrature rule presented in Velázquez-Mata et al. [1] to the hypersingular BEM formulation, enabling an accurate computation of integrals with higher order singularities.

The paper is organised as follows. First, the hypersingular BEM formulation is briefly described, and the numerical quadrature is developed to consider singularities of type $1/r^2$ by evaluating the corresponding generalised moment. Then, the proposed methodology is numerically verified by solving the Helmholtz equation on an open boundary. Dirichlet and Neumann boundary conditions are both prescribed, and the results computed by the classical and hypersingular BEM formulations are compared. Second, an illustration of the applicability of the proposed technique for solving the Helmholtz equation in domains with inclusions is presented. The BEM and TBEM formulations are both coupled to represent null-thickness boundaries. Finally, the main results of this research are summarised.

2 BEM FORMULATIONS

The starting point for the BEM formulation is the BIE (eqn (1)). Once the boundary is discretised into N elements, $\Gamma = \bigcup_{j=1}^{N} \Gamma^j$, and the field variables within an element Γ^j are approximated from the nodal values u^i and t^i through the element shape functions $\phi^i(\mathbf{x})$ of order p, the eqn (1) is rewritten as follows:

$$c(\mathbf{x}^*)u(\mathbf{x}^*) = \sum_{j=1}^{N} \sum_{i=0}^{p} \left[\left(\int_{\Gamma^j} \phi^i(\mathbf{x}) \mathcal{G}(\mathbf{x}, \mathbf{x}^*) \, d\Gamma \right) t^i - \left(\int_{\Gamma^j} \phi^i(\mathbf{x}) \mathcal{H}(\mathbf{x}, \mathbf{x}^*) \, d\Gamma \right) u^i \right] \quad (3)$$

Similarly, the hypersingular form of the BIE (eqn (2)) leads to the following expression after boundary discretisation:

$$c(\mathbf{x}^*)t(\mathbf{x}^*) = \sum_{j=1}^{N} \sum_{i=0}^{p} \left[\left(\int_{\Gamma^j} \phi^i(\mathbf{x}) \tilde{\mathcal{G}}(\mathbf{x}, \mathbf{x}^*) \, d\Gamma \right) t^i - \left(\int_{\Gamma^j} \phi^i(\mathbf{x}) \tilde{\mathcal{H}}(\mathbf{x}, \mathbf{x}^*) \, d\Gamma \right) u^i \right] \quad (4)$$

Eqns (3) and (4), can be coupled to produce the Dual BEM formulation, as will be done in Section 4.

A critical step in the BEM formulation is the integration of elements, as mentioned above. The integral kernel usually has a singularity that depends on the physical problem. Moreover, the kernels of the hypersingular formulation are strongly singular. Typically, these singularities are of the form $\log(r)$, $1/r$ and $1/r^2$. The element integration should then be computed according to the singularity but also to the element order p used for the field approximation.

It was shown in Romero et al. [4] that the element shape functions $\phi^i(\mathbf{x})$ of order p based on a Lagrange interpolant can be derived from the Bernstein basis as:

$$\phi^i(t) = \sum_{k=0}^{n} c_k^i B_k^n(t), \quad i = 0, \ldots, p \quad (5)$$

where c_k^i are control points and $B_k^n(t)$ is the Bernstein polynomial of order n defined over the interval $t \in [0, 1]$:

$$B_k^n(t) = \binom{n}{k} t^k (1-t)^{n-k}, \quad k = 0, \ldots, n \quad (6)$$

The Lagrange interpolant must fulfil $\phi^i(t_j) = \sum_{k=0}^{n} c_k^i B_k^n(t_j) = \delta_{ij}$, $j = 0, \ldots, n$ at the element nodes t_j, where δ_{ij} is the Kronecker delta. This condition is commonly expressed as a linear system of equations through the Bernstein–Vandermonde matrix $A_{jk} = B_k^n(t_j)$.

Once the control points c_k^i are obtained, the element integration in eqns (3) and (4) can be rewritten as follows in the natural coordinate ξ according to eqn (5):

$$
\int_{-1}^{1} \phi^i(\xi) \mathcal{F}(\xi, \mathbf{x}^*) \frac{d\Gamma}{d\xi} \, d\xi = \int_{0}^{1} \phi^i(t) \mathcal{F}(t, \mathbf{x}^*) \frac{d\Gamma}{d\xi} \frac{d\xi}{dt} \, dt
$$

$$
= \int_{0}^{1} \left(\sum_{k=0}^{n} c_k^i B_k^n(t) \right) \mathcal{F}(t, \mathbf{x}^*) \frac{d\Gamma}{d\xi} \frac{d\xi}{dt} \, dt \tag{7}
$$

$$
= \sum_{k=0}^{n} c_k^i \left(\int_{0}^{1} B_k^n(t) \mathcal{F}(t, \mathbf{x}^*) \frac{d\Gamma}{d\xi} \frac{d\xi}{dt} \, dt \right)
$$

where, $\mathcal{F}(t, \mathbf{x}^*)$ stands for the type of singularity in the fundamental solution.

The quadrature rule presented by the authors in Velázquez-Mata et al. [1] is extended in this work to assess hypersingular kernels, as well as weakly singular and singular integrations. The proposed quadrature allows us to evaluate the BIE numerically for any element order p.

3 QUADRATURE RULES

The quadrature rules proposed here are based on the recent work published by Velázquez-Mata et al. [1]. However, one of the novelties of this paper is to develop a numerical quadratures so that integrals with singularities of the type $1/r^2$ can be computed without loss of accuracy in the estimation of weakly singular and singular integrals. Thus, the quadrature rule enable the calculation of eqn (7) accounting for $\mathcal{F}(\xi, \mathbf{x}^*)$ equals 1, $\log|\xi^* - \xi|$, $(\xi^* - \xi)^{-1}$ and $(\xi^* - \xi)^{-2}$, where ξ^* is the natural coordinate of the collocation point at the integration element.

The quadrature of order M should approximate the integrals in eqn (7) as:

$$
\int_{-1}^{1} B_k^n(\xi) \mathcal{F}(\xi, \mathbf{x}^*) \, d\xi = \int_{0}^{1} B_k^n(t) \mathcal{F}(t, \mathbf{x}^*) \frac{d\xi}{dt} \, dt \simeq \sum_{m=0}^{M} B_k^n(t_m) \mathcal{F}(t_m, \mathbf{x}^*) \frac{d\xi}{dt} w_m \tag{8}
$$

where t_m and w_m are the integration points and weights, respectively. Quadrature weights are obtained from the solution of a system of equations defined from the above approximation [1]:

$$
\sum_{m=0}^{M} \psi_k(t_m, \xi^*) w_m = m_k, \qquad k = 0, \ldots, n \tag{9}
$$

where $\psi_k(t_m, \xi^*) = B_k^n(t) \mathcal{F}(t_m, \mathbf{x}^*) d\xi/dt$ stands for the integral kernel and m_k represents the generalised moment, that is, the exact solution of eqn (8). The generalised moments can be obtained from the Brandaõ approach to the finite part integrals [5] according to Carley [6]. The solution of the generalised moment for weakly singular and singular integrals can be found in Velázquez-Mata et al. [1].

The following generalised moment should be included in the methodology for integrating hypersingular kernels:

$$
m_k = \mathrm{FP} \int_{-1}^{1} \frac{B_k^n(\xi)}{(\xi^* - \xi)^2} \, d\xi = \mathrm{FP} \int_{0}^{1} \frac{B_k^n(t)}{(\xi^* - 2t + 1)^2} \frac{d\xi}{dt} \, dt \tag{10}
$$

In addition to the generalised moment proposed in Velázquez-Mata et al. [1], the solution of the generalised moment expressed on the univariate basis $t \in [0, 1]$ is obtained according to the following formulas [5], [6]:

$$
\begin{aligned}
m_k &= \text{FP} \int_0^1 \frac{B_k^n(t)}{(\xi^* - 2t + 1)^2} \frac{d\xi}{dt} \, dt \\
&= B_k^n(t^*) \text{FP} \int_0^1 \frac{1}{(\xi^* - 2t + 1)^2} \frac{d\xi}{dt} \, dt - \frac{dB_k^n(t^*)}{dt} \frac{dt}{d\xi} \text{CPV} \int_0^1 \frac{1}{\xi^* - 2t + 1} \frac{d\xi}{dt} \, dt \quad (11) \\
&\quad + \int_0^1 \frac{B_k^n(t) - B_k^n(t^*) + \frac{dB_k^n(t^*)}{dt} \frac{dt}{d\xi}(\xi^* - 2t + 1)}{\xi^* - 2t + 1} \frac{d\xi}{dt} \, dt
\end{aligned}
$$

where

$$
\text{FP} \int_0^1 \frac{1}{(\xi^* - 2t + 1)^2} \frac{d\xi}{dt} \, dt = \begin{cases} 2/(\xi^{*2} - 1) & |\xi^*| \neq 1 \\ -1/2 & \xi^* = \pm 1 \end{cases} \quad (12)
$$

$$
\text{CPV} \int_0^1 \frac{1}{\xi^* - 2t + 1} \frac{d\xi}{dt} \, dt = \begin{cases} \log\left|\frac{\xi^* + 1}{1 - \xi^*}\right| & |\xi^*| \neq 1 \\ \pm \log(2) & \xi^* = \pm 1 \end{cases} \quad (13)
$$

Then, the system of eqn (9) includes $n + 1$ equations for each type of function to be integrated. The quadrature rule proposed in this work is capable of integrating kernels with constant, $\log(r)$, $1/r$, and $1/r^2$ terms when the collocation point belongs to the integration element, which are found in the fundamental solution or in their series expansions. Therefore, eqn (9) defines a system of $4(n + 1)$ equations with $M + 1$ unknown weights w_m. The solution is obtained in the least-squares sense when overdetermined and in the minimum norm least-squares sense when undetermined. The integration point should not be located at the natural coordinates of the collocation nodes to avoid indeterminate terms in eqn (9).

3.1 Numerical validation

Table 1 summarises the integral values calculated from $\text{FP} \int_{-1}^1 \phi^i (\xi - \xi^*)^{-2} \, d\xi$ for the five shape functions corresponding to the Chebyshev points of the first kind, of order $p = 4$ and $\xi^* = 0$. This integrand presents a singularity at $\xi = 0$ for the shape function $i = 2$ with nonzero value at ξ^*. The integral is numerically evaluated using the proposed approach and using the built-in MATLAB function integral [7] for comparison purposes, and both solutions are compared with the exact values computed from eqns (7) and (11). The integral is only correctly evaluated when the proposed methodology is used.

Fig. 1 shows the L_2 scaled error ϵ_2 in the integral computation in eqn (10) for different element order and point distributions: (i) Chebyshev points of the first kind; (ii) Chebyshev points of the second kind; (iii) LGL integration points; and (iv) equidistant nodes [4]. The accuracy of the proposed quadrature rules is analysed for different numbers of integration points M according to the shape function of order p. The quadrature rule gave errors lower than $\mathcal{O}(10^{-5})$ in all cases. The integration error increases with the element order p, and is slightly affected by the number of integration points for $M \geq 4(p + 1)$. Therefore, a value for M equal to $4(p + 1)$ is considered, as in Velázquez-Mata et al. [1] for singularities of types $\log(r)$ and $1/r$.

Moreover, an additional test was performed as that presented in Guiggiani and Casalini [8] by Guiggiani and Casalini to calculate the finite part of the function $g(x) = (x^3 + 1)/x^2$. Two

Table 1: Computed integral values of FP $\int_{-1}^{1} \phi^i (\xi - \xi^*)^{-2} \, d\xi$ for shape function ϕ^i of order $p = 4$ defined at Chebyshev points of the first kind and $\xi^* = 0$: (i) evaluated by the proposed approach (Q), (ii) using the built-in `integral` MATLAB function (I), and (iii) exact values (eqns (7) and (11)) (M). The values corresponding to the shape function that presents a non-zero value at $\xi = \xi^*$ are highlighted in grey.

i	M	Q	I
0	-0.024045	-0.024045	-0.023959
1	2.957379	2.957379	2.957178
2	-7.866667	-7.866667	312551427532.072510
3	2.957379	2.957379	2.957332
4	-0.024045	-0.024045	-0.024059

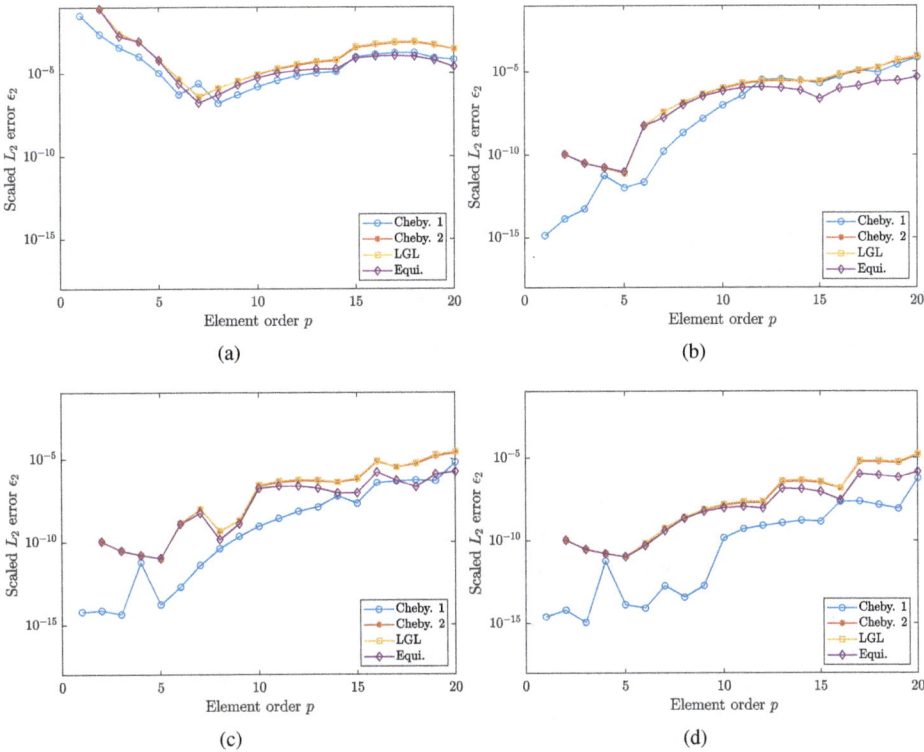

Figure 1: L_2 scaled error ϵ_2 in the computation of the integral in eqn (10) using (a) $M = 2(p+1)$; (b) $M = 3(p+1)$; (c) $M = 4(p+1)$; and (d) $M = 8(p+1)$ integration points.

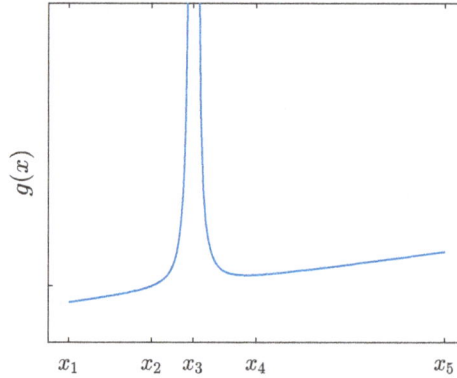

Figure 2: Test on the integration procedure.

quadratic elements $\Gamma^a := [x_1, x_2, x_3]$ and $\Gamma^b := [x_3, x_4, x_5]$ were used to integrate. The test consists of the calculation of the finite part given by:

$$\text{FP} \int_{x_2}^{x_4} \frac{x^3 + 1}{x^2} dx \tag{14}$$

The integrand has singularity at $x = 0$ of order $1/r^2$ (see Fig. 2). The exact value of the finite part is [9]:

$$\text{FP} \int_{x_2}^{x_4} \frac{x^3 + 1}{x^2} dx = \frac{x_4^2 - x_2^2}{2} - \frac{1}{x_4} + \frac{1}{x_2} \tag{15}$$

The finite part of the integral was computed for the following nodal coordinates: $x_1 = \{-1.5, -2, -3\}$, $x_2 = -1$, $x_3 = 0$, $x_4 = 1.5$, and $x_5 = \{3, 4, 6\}$, being the exact value -1.041667. The nodal coordinates produce distorted elements. Therefore, the relation between x and ξ established by the shape functions in eqns (16) and (17) is non-linear.

$$x(\xi) = \phi^1(\xi)x_1 + \phi^2(\xi)x_2 + \phi^3(\xi)x_3 \quad \text{on} \quad \Gamma^a \tag{16}$$

$$x(\xi) = \phi^1(\xi)x_3 + \phi^2(\xi)x_4 + \phi^3(\xi)x_5 \quad \text{on} \quad \Gamma^b \tag{17}$$

It should be indicated that the locations of x_1 and x_5 do not affect the value of the integral, only the transformation between x and ξ. In these cases, the finite part integral could be not invariant with respect to the transformation, and this should be considered in the computations [10]. Table 2 summarised the computed results. In all tests, the error, with respect to the exact value, was lower than 1×10^{-9}.

4 CASE STUDY

In this section, a rectangular fluid duct defined by the domain $\Omega := [0, 1.225] \times [-0.1, 0.1]$ with flat surfaces is studied. The fluid, with density $\rho = 1.225 \, \text{kg/m}^3$, allows dilatational waves to propagate with a velocity of $c_f = 340 \, \text{m/s}$. The boundary was defined as follows:

$$\Gamma_1 := [0, 1.225] \times [-0.1, -0.1], \Gamma_2 := [1.225, 1.225] \times [-0.1, 0.1],$$
$$\Gamma_3 := [1.225, 0] \times [0.1, 0.1], \Gamma_4 := [0, 0] \times [0.1, -0.1]. \tag{18}$$

The boundary Γ_4 was subjected to a uniform normal velocity $v_n^i = 1 \, \text{m/s}$, while the opposite boundary Γ_2 had a non-reflecting condition given by $p^i/v_n^i = \rho c_f$. The boundary conditions were the same as set $v_n^i = 0$ at $\Gamma_{1,3}$. The problem wavelength was defined from the duct

Table 2: Results of the test on the integration procedure.

Case	x_1	x_2	x_3	x_4	x_5	Exact value	Numerical value	Error [$\times 10^{-9}$]
1	−1.5	−1.0	0.0	1.5	3.0	1.041667	1.041667	0.218811
2	−2.0	−1.0	0.0	1.5	3.0	1.041667	1.041667	0.181570
3	−3.0	−1.0	0.0	1.5	3.0	1.041667	1.041667	0.144573
4	−1.5	−1.0	0.0	1.5	4.0	1.041667	1.041667	0.201584
5	−2.0	−1.0	0.0	1.5	4.0	1.041667	1.041667	0.164343
6	−3.0	−1.0	0.0	1.5	4.0	1.041667	1.041667	0.127347
7	−1.5	−1.0	0.0	1.5	6.0	1.041667	1.041667	0.184385
8	−2.0	−1.0	0.0	1.5	6.0	1.041667	1.041667	0.147145
9	−3.0	−1.0	0.0	1.5	6.0	1.041667	1.041667	0.110148

length ($L = 1.225$ m), being $\lambda = L/20$ m, leading to a frequency of 5551 Hz. The model was discretised into elements with nodes located at Chebyshev points of the first kind of length h ensuring that $\kappa h = 3$, and order $p = 6$. This problem has an analytical solution for the one-dimensional case [11] that can be used for comparison purposes, given the geometry of the problem.

Fig. 3 shows a comparison between the analytical solution of the problem and the numerical result obtained using both the BEM formulation and the hypersingular one.

Once the good performance of the formulation had been verified, a straight inclusion in the duct was considered to work as a barrier. Fig. 4 schematises the geometry of the duct and inclusion. The inclusion in the duct has a small thickness equal to 10^{-4} m and can therefore be considered null thickness. In this case, eqns (3) and (4) should be coupled to obtain the unknowns of the problem using the dual BEM.

Eqn (3) led to the following equation when the collocation point $\mathbf{x}^*$ belongs to the duct boundary:

$$c(\mathbf{x}^*)u(\mathbf{x}^*) = \sum_{j=1}^{N_1}\sum_{i=0}^{p}\left[\left(\int_{\Gamma^j}\phi^i(\mathbf{x})\mathcal{G}(\mathbf{x},\mathbf{x}^*)\,d\Gamma\right)t^i - \left(\int_{\Gamma^j}\phi^i(\mathbf{x})\mathcal{H}(\mathbf{x},\mathbf{x}^*)\,d\Gamma\right)u^i\right] +$$

$$\sum_{j=1}^{N_2}\sum_{i=0}^{p}\left[\left(\int_{\Gamma^j}\phi^i(\mathbf{x})\mathcal{G}(\mathbf{x},\mathbf{x}^*)\,d\Gamma\right)(t^{i+}+t^{i-}) - \left(\int_{\Gamma^j}\phi^i(\mathbf{x})\mathcal{H}(\mathbf{x},\mathbf{x}^*)\,d\Gamma\right)(u^{i+}-u^{i-})\right]$$

$$(19)$$

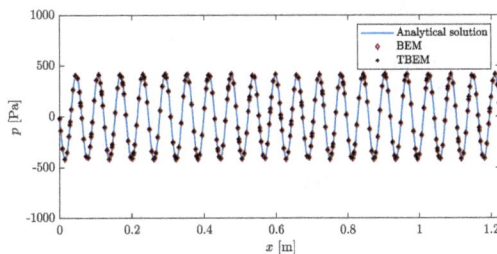

Figure 3: Analytical solution and numerical results in the middle line ($y = 0$) of a channel with height $h_d = 0.2$ m.

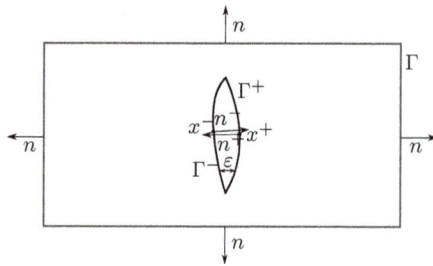

Figure 4: Geometry of the duct and inclusion ($\varepsilon \to 0$).

where N_1 and N_2 are the elements that represent the duct and inclusion with normal n^+, respectively; and u^i and t^i are related to the pressure and normal velocity, respectively.

In eqn (19), the following relations have been taken into account:

$$\int_{\Gamma^j} \phi^i(\mathbf{x}^+)\mathcal{G}(\mathbf{x}^+, \mathbf{x}^*)\, d\Gamma = \int_{\Gamma^j} \phi^i(\mathbf{x}^-)\mathcal{G}(\mathbf{x}^-, \mathbf{x}^*)\, d\Gamma \tag{20}$$

$$\int_{\Gamma^j} \phi^i(\mathbf{x}^+)\mathcal{H}(\mathbf{x}^+, \mathbf{x}^*)\, d\Gamma = -\int_{\Gamma^j} \phi^i(\mathbf{x}^-)\mathcal{H}(\mathbf{x}^-, \mathbf{x}^*)\, d\Gamma \tag{21}$$

corresponding to points $\mathbf{x}^+$ and $\mathbf{x}^-$ at the same barrier location but with normals n^+ and n^-, respectively.

Eqn (3) can also be applied for $\mathbf{x}^* = \mathbf{x}^+$ and $\mathbf{x}^* = \mathbf{x}^-$ leading to two equations that can be added as follows:

$$\frac{1}{2}\left(u(\mathbf{x}^+) + u(\mathbf{x}^-)\right) = \sum_{j=1}^{N_1}\sum_{i=0}^{p}\left[\left(\int_{\Gamma^j}\phi^i(\mathbf{x})\mathcal{G}(\mathbf{x},\mathbf{x}^+)\, d\Gamma\right)t^i - \left(\int_{\Gamma^j}\phi^i(\mathbf{x})\mathcal{H}(\mathbf{x},\mathbf{x}^+)\, d\Gamma\right)u^i\right]$$
$$+ \sum_{j=1}^{N_2}\sum_{i=0}^{p}\left[\left(\int_{\Gamma^j}\phi^i(\mathbf{x})\mathcal{G}(\mathbf{x},\mathbf{x}^+)\, d\Gamma\right)(t^{i+} + t^{i-})\right.$$
$$\left. - \left(\int_{\Gamma^j}\phi^i(\mathbf{x})\mathcal{H}(\mathbf{x},\mathbf{x}^+)\, d\Gamma\right)(u^{i+} - u^{i-})\right] \tag{22}$$

Similarly, eqn (4) allows us to obtain:

$$\frac{1}{2}\left(t(\mathbf{x}^+) - t(\mathbf{x}^-)\right) = \sum_{j=1}^{N_1}\sum_{i=0}^{p}\left[\left(\int_{\Gamma^j}\phi^i(\mathbf{x})\tilde{\mathcal{G}}(\mathbf{x},\mathbf{x}^+)\, d\Gamma\right)t^i - \left(\int_{\Gamma^j}\phi^i(\mathbf{x})\tilde{\mathcal{H}}(\mathbf{x},\mathbf{x}^+)\, d\Gamma\right)u^i\right]$$
$$+ \sum_{j=1}^{N_2}\sum_{i=0}^{p}\left[\left(\int_{\Gamma^j}\phi^i(\mathbf{x})\tilde{\mathcal{G}}(\mathbf{x},\mathbf{x}^+)\, d\Gamma\right)(t^{i+} + t^{i-})\right.$$
$$\left. - \left(\int_{\Gamma^j}\phi^i(\mathbf{x})\tilde{\mathcal{H}}(\mathbf{x},\mathbf{x}^+)\, d\Gamma\right)(u^{i+} - u^{i-})\right] \tag{23}$$

Eqns (19), (22) and (23) led to the dual BEM equation system that can be used to compute the pressure and the normal velocity in the duct (u^i and t^i), and at inclusion (u^{i+}, u^{i-}, t^{i+}, and t^{i-}), once the boundary conditions are imposed.

With this procedure, three problems for different-sized straight inclusions were solved. Using the geometry of the verification duct, two new coincident patches with normals in opposite directions were added in each case. Placed in the middle of the channel, the lengths of the inclusions were considered as a percentage of the height of the duct h_d: $0.3\,h_d$, $0.6\,h_d$ and $0.9\,h_d$, respectively. Thus, the two new patches implemented in each problem are as follows:

$$\Gamma_5 := [0.6125, 0.6125] \times [0.03, -0.03], \Gamma_6 := [0.6125, 0.6125] \times [-0.03, 0.03] \quad (24a)$$
$$\Gamma_5 := [0.6125, 0.6125] \times [0.06, -0.06], \Gamma_6 := [0.6125, 0.6125] \times [-0.06, 0.06] \quad (24b)$$
$$\Gamma_5 := [0.6125, 0.6125] \times [0.09, -0.09], \Gamma_6 := [0.6125, 0.6125] \times [-0.09, 0.09] \quad (24c)$$

The fluid properties, as well as the boundary conditions set in the duct, the wavenumber value, and the element discretization remained the same as in the non-barrier problem. Meanwhile, for $\Gamma_{5,6}$ the normal velocity was imposed as $v_n^i = 0$.

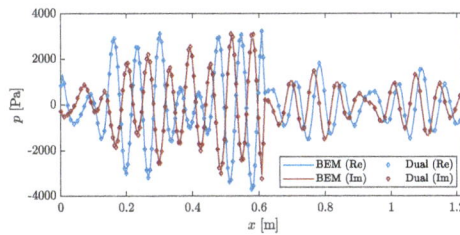

(a)

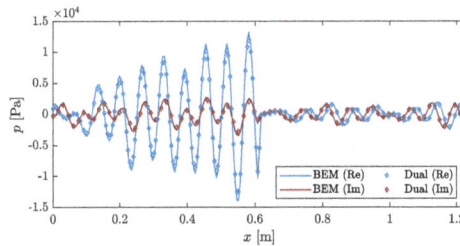

(b)

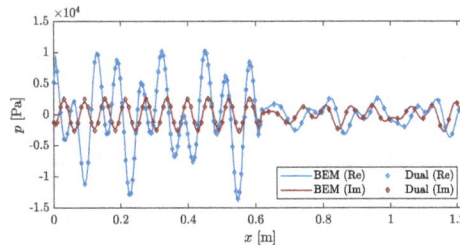

(c)

Figure 5: Pressure values for a frequency of 5551 Hz in the middle line ($y = 0$) of a channel of height $h_d = 0.2\,$m. Comparison between the numerical results obtained with the two numerical formulations with a barrier length of: (a) $0.3\,h_d$; (b) $0.6\,h_d$; and (c) $0.9\,h_d$.

The main advantage of the proposed Dual BEM procedure lies in the possibility of easily introducing thin elements, achieving accurate results through the use of fewer patches.

Fig. 5 shows the pressure in the middle line ($y = 0$) of the three problems considered. Each graph displays a comparison between the results obtained by using the BEM formulation (solid lines) and Dual BEM formulation (markers) for both imaginary and real values. The good agreement between the curves in each case is easily noticeable, which confirms the strong performance of the proposed formulation.

5 CONCLUSIONS

This paper enhances the abilities of the QUEEN quadrature presented in Velázquez-Mata et al. [1] to solve engineering problems. First, the numerical quadrature has been extended to consider singularities of type $1/r^2$, since the original quadrature only enables computation of singularities such as $\log(r)$ and $1/r$. This novelty is highly successful, to the best of the author's knowledge, in developing BEM codes without the need for the regularisation process. Second, the novel quadrature has been used to solve the Helmholtz equation on a boundary with open geometry. The results of the classical and hypersingular BEM formulations have been shown, highlighting the capabilities of each methodology. The hypersingular BEM formulation was more accurate for computing the considered problems where Dirichlet boundary conditions were prescribed, while the classical BEM formulation was more accurate in the case of Neumann boundary conditions.

Finally, a more complex problem consisting of a rectangular duct that includes a small-thickness barrier has been analysed to show the capabilities of the quadrature in the Dual BEM formulation.

ACKNOWLEDGEMENTS

The authors would like to acknowledge the financial support provided by the Spanish Ministry of Science and Innovation under the research project PID2019-109622RB-C21, and US-126491 funded by the FEDER Andalucía 20142020 Operational Program. The support of the Andalusian Scientific Computing Centre (CICA) is grateful.

REFERENCES

[1] Velázquez-Mata, R., Romero, A., Domínguez, J., Tadeu, A. & Galvín, P., A novel high-performance quadrature rule for BEM formulations. *Engineering Analysis with Boundary Elements*, **140**, pp. 607–617, 2022.

[2] Domínguez, J., *Boundary Elements in Dynamics*. Computational Mechanics Publications and Elsevier Applied Science, 1993.

[3] Guiggiani, M., *Formulation and Numerical Treatment of Boundary Integral Equations with Hypersingular Kernels*. Computational Mechanics Publications: Southampton and Boston, 1998.

[4] Romero, A., Galvín, P., Cámara-Molina, J. & Tadeu, A., On the formulation of a BEM in the Bézier–Bernstein space for the solution of Helmholtz equation. *Applied Mathematical Modelling*, **74**, pp. 301–319, 2019.

[5] Brandaõ, M.P., Improper integrals in theoretical aerodynamics: The problem revisited. *AIAA Journal*, **25**(9), pp. 1258–1260, 1987.

[6] Carley, M., Numerical quadratures for singular and hypersingular integrals in boundary element methods. *SIAM Journal on Scientific Computing*, **29**(3), pp. 1207–1216, 2007.

[7] MATLAB, version 9.11.0.1837725 (R2021b). The MathWorks Inc.: Natick, MA, 2021.

[8] Guiggiani, M. & Casalini, P., Direct computation of cauchy principal value integrals in advanced boundary elements. *International Journal for Numerical Methods in Engineering*, **24**(9), pp. 1711–1720, 1987.

[9] Kaya, A.C. & Erdogan, F., On the solution of integral equations with strongly singular kernels. *Quarterly of Applied Mathematics*, **45**(1), pp. 105–122, 1987.

[10] Kutt, H.R., *On the Numerical Evaluation of Finite-Part Integrals Involving an Algebraic Singularity*. Stellenbosch University, 1975.

[11] Biermann, J., von Estorff, O., Petersen, S. & Wenterodt, C., Higher order finite and infinite elements for the solution of helmholtz problems. *Computer Methods in Applied Mechanics and Engineering*, **198**(13), pp. 1171–1188, 2009.

REAL- AND COMPLEX-VARIABLE IMPLEMENTATIONS OF THE CONSISTENT BOUNDARY ELEMENT METHOD IN TWO-DIMENSIONAL ELASTICITY: A COMPARATIVE ASSESSMENT

NEY AUGUSTO DUMONT

Department of Civil and Environmental Engineering, Pontifical Catholic University of Rio de Janeiro, Brazil

ABSTRACT

The collocation boundary element method has recently been entirely revisited by the author. Arbitrary rigid-body displacements, as for elasticity, are naturally taken into account, and traction force parameters are always in balance independently of problem scale and mesh discretization. For generally curved boundaries, the correct definition of traction force interpolation functions enables the enunciation of a general convergence theorem, the introduction of patch and cut-out tests, and, not least, a considerable simplification of the numerical implementations. Simple code schemes for the 2D formulation are proposed exclusively in terms of Gauss–Legendre quadrature for arbitrarily high – actually only machine-precision dependent – computational accuracy of results independently of a problem's geometry and topology. On the other hand, the complex-variable formulation of the problem leads to more simplicity of implementation and numerical results that seem less liable to round-off errors. We propose in this contribution the comparative assessment of the real- and complex-variable formulations regarding coding efficiency, computational effort, error estimation, and numerical precision and accuracy of results for applications that are topologically highly challenging, with source-field distances in the subnanometer range.

Keywords: boundary elements, consistent formulation, machine-precision integration, complex-variable formulation.

1 INTRODUCTION

We have recently proposed the entire reformulation of the collocation boundary element method (CBEM), which keeps its essence, as obtainable from the application of Somigliana's identity, but introduces paramount features that go back to the initial consistency concerns related to the variationally, hybrid boundary element method [1], incorporates concepts related to the adequate treatment of quasi-singularities for 2D problems [2], [3], revisits the formulation in terms of spectral consistency assessments [4], [5] and ends up with the introduction of important code-implementation improvements [6]. All these developments have culminated in the comprehensive proposition of the CBEM [7], in general, with 2D developments laid down in Dumont [8], [9]. Although we have introduced in Dumont [2] and Dumont and Noronha [3] the paradigm-shifting concept of complex quasi-singularity poles (in terms of the boundary parametric variable ξ that may eventually assume values $\xi_s = a \pm ib$ related to a general source point s), all developments had been proposed in terms of Cartesian real variables (x, y), for 2D problems. Quite recently, we have envisaged that the complex-variable reformulation of Dumont [8], thus in terms of $z = x + iy$, may eventually lead to overall simplification of the implemented code and contribute to meliorate the unavoidable round-off-error issues that arise in applications to extremely challenging topologies [10].

This short contribution conveys the comparative assessment of the real-variable formulation of Dumont [8] and its complex-variable counterpart proposed in Dumont [10]. Owing to space restrictions we only address the evaluation of stress results at internal

WIT Transactions on Engineering Sciences, Vol 135, © 2023 WIT Press
www.witpress.com, ISSN 1743-3533 (on-line)
doi:10.2495/BE460061

points, which is actually the most critical issue one may face particularly when dealing with extremely small source-to-field distances. Interested readers are referred to the abundant details given in Dumont [8]–[10].

2 BASIC PROBLEM OUTLINE

Whether using real or complex variables, the basic system matrix to be solved in the frame of the CBEM has the format

$$\mathbf{H}(\mathbf{d} - \mathbf{d}^p) = \mathbf{G}(\mathbf{t} - \mathbf{t}^p)_{ad} \tag{1}$$

In this equation, $\mathbf{H}$ is the square, double-layer potential matrix of order $n_d = 2n_n$, for 2D elasticity and the problem discretized with n_n nodal points, and $\mathbf{G}$ is the single-layer potential matrix with n_d rows and $n_d + 2n_e$ columns, as we code for n_e elements of any order o_e, in principle taking into account that the left and right tangents at a nodal point connecting two elements are different. As laid down in Dumont [5], [7], we are assuming just for the sake of elegant and compact formulation that some *particular* solution of interest is known – whether or not related to non-zero body forces – and may be approximately expressed as boundary nodal displacement $\mathbf{d}^p$ and traction $\mathbf{t}^p$ data. The problem's primary boundary displacement and traction parameters are $\mathbf{d}$ and $\mathbf{t}$, which are in part known and in part to be obtained in the frame of a general mixed-boundary formulation. As comprehensively assessed in Dumont [4], [7]–[9], we write for consistency that the traction $(\mathbf{t} - \mathbf{t}^p)_{ad}$ is *admissible*, in equilibrium with the applied domain forces: this follows the same mathematical/mechanical principle that, since for a finite domain rigid-body displacement amounts of $(\mathbf{d} - \mathbf{d}^p)$ cannot be transformed into forces, also non-equilibrated forces should not be transformed into displacements.

3 EVALUATION OF STRESS RESULTS AT INTERNAL POINTS

3.1 Real-variable formulation

We follow the notation given in Dumont [8] to represent stress results at an internal point s according to the expression given, for instance, in Brebbia et al. [11], although with some index transposition while always keeping in mind the traction-force interpolation given by eqn (20) of Dumont [7], for integration carried out along successive boundary segments of Γ in terms of the parametric variable $\xi \in [0, 1]$:

$$\sigma_{is} = \sigma_{is}^p + |J|_{(at\ \ell)} \int_{\Gamma} u_{is\ell}^* N_{(at\ \ell)}^{o_e} \, \mathrm{d}\xi \, (t_{\ell} - t_{\ell}^p)_{ad}$$
$$- \int_{\Gamma} p_{isn}^* N_{(at\ n)}^{o_e} |J| \, \mathrm{d}\xi \, (d_n - d_n^p) \tag{2}$$

In this expression, σ_{is} is the symmetric stress tensor and

$$u_{is\ell}^* \equiv -\sigma_{is\ell}^* = \frac{(1-2\nu)(r_{,i}\delta_{s\ell} + r_{,s}\delta_{i\ell} - r_{,\ell}\delta_{is}) + 2r_{,\ell}r_{,i}r_{,s}}{4\pi(1-\nu)r} \tag{3}$$

$$p_{isn}^* = G\frac{(1-2\nu)r_{,n}\delta_{si} + \nu(\delta_{ns}r_{,i} + \delta_{ni}r_{,s}) - 4r_{,i}r_{,s}r_{,n}}{\pi(1-\nu)r^2}\frac{\partial r}{\partial n}$$
$$+ \frac{G}{2\pi(1-\nu)}\left\{2\nu\left(r_{,i}r_{,n}n_s + r_{,s}r_{,n}n_i\right) - (1-4\nu)\delta_{si}n_n\right.$$
$$\left. + (1-2\nu)(2r_{,i}r_{,s}n_n + \delta_{ni}n_s + \delta_{ns}n_i)\right\} \tag{4}$$

In these equations, repeated indices mean summation and a comma (,) in the subscript means derivative with respect to the corresponding Cartesian direction 1 or 2, referring to x or y. The unit normal to the boundary is characterized by n and its Cartesian projections are also given. The Kronecker delta is given as δ_{si}, for instance. Boundary displacement and traction are approximated along successive segments by the scalars $N^{o_e}_{(at\ n)}(\xi)$ and $|J_{(at\ \ell)}|N^{o_e}_{(at\ \ell)}(\xi)/|J(\xi)|$, in terms of known *nodal* $(d_n - d_n^p)$ and *locus* $(t_\ell - t_\ell^p)_{ad}$ data – according to the important, corrected, formulation of the CBEM proposed in Dumont [5], [7]. The superscript o_e stands for the polynomial interpolation order, here implemented for linear, quadratic, cubic and quartic elements. Since repeated indices mean summation, we use $(at\ n)$ and $(at\ \ell)$ in eqn (2) to indicate that the shape functions refer to node n and locus ℓ [7]. The interpolation functions $N^{o_e}_{(at\ n)}$ and $N^{o_e}_{(at\ \ell)}$ have local support: the boundary integration along Γ prescinds the specific indication of sum over segments. The tractions are assumed to be *admissible*, that is, in balance with applied domain forces (whether void or not), which is the reason of the assigned subscript $(\ )_{ad}$ [4], [7]. Moreover, as proposed in Dumont [7], [8], we already take into account eventual *particular* or body-force solutions, characterized by the superscript $(\ )^p$, whenever this is mathematically feasible [7].

In the code implementation of Dumont [8], we express the stress tensor as a vector $\boldsymbol{\sigma} \equiv \langle \sigma_x\ \sigma_y\ \tau_{xy} \rangle^{\mathrm{T}}$, and $u^*_{is\ell}$ and p^*_{isn} as arrays

$$u^*_{is\ell} = \frac{1}{2\pi(1-\nu)} \left\{ \frac{0.5-\nu}{r^2} \begin{bmatrix} x & -y \\ -x & y \\ y & x \end{bmatrix} + \frac{1}{r^4} \begin{bmatrix} x^3 & x^2 y \\ xy^2 & y^3 \\ x^2 y & xy^2 \end{bmatrix} \right\} \tag{5}$$

$$p^*_{isn} = \frac{G}{2\pi(1-\nu)|J|} \left\{ \frac{1}{r^2} \begin{bmatrix} y' & -x' \\ y' & -x' \\ -x' & y' \end{bmatrix} + \frac{1}{r^4} \begin{bmatrix} 4x^2 y' - 2xyx' & 2xyy' \\ -2xyx' & -4y^2 x' + 2xyy' \\ 2xyy' & -2xyx' \end{bmatrix} \right.$$

$$\left. - \frac{8(xy' - yx')}{r^6} \begin{bmatrix} x^3 & x^2 y \\ xy^2 & y^3 \\ x^2 y & xy^2 \end{bmatrix} \right\} \tag{6}$$

where, quoting [8], "$n_x = y'/|J|$ and $n_y = -x'/|J|$. In these expressions, the rows correspond to the elements of the stress vector defined above and the columns to the Cartesian directions represented by s. The subscripts ℓ and n are implicit as the reference to the field points related to $N^{o_e}_{(at\ \ell)}$ and $N^{o_e}_{(at\ n)}$ for the function multiplications and eventual integrations indicated by eqn (2). Moreover, the matrix components have been manipulated in order to make the quasi-singularity terms $1/r^2$, $1/r^4$ and $1/r^6$ explicit (they do not necessarily reflect the actual quasi-singularity powers), which is paramount for the subsequent developments."

3.2 Complex-variable formulation

It is proposed in Dumont [10] the complex-variable expression for stress results at internal points

$$\begin{Bmatrix} \sigma_x + \sigma_y \\ \sigma_y - \sigma_x + 2i\tau_{xy} \end{Bmatrix}_s = \begin{Bmatrix} \sigma_x^p + \sigma_y^p \\ \sigma_y^p - \sigma_x^p + 2i\tau_{xy}^p \end{Bmatrix}_s$$

$$+ \frac{|J|_{(at\ \ell)}}{4\pi(1-\nu)} \int_\Gamma N^{o_e}_\ell \begin{bmatrix} \dfrac{1}{z} & \dfrac{1}{\bar{z}} \\ -\dfrac{\bar{z}}{z^2} & \dfrac{3-4\nu}{z} \end{bmatrix} \mathrm{d}\xi \begin{Bmatrix} (t - t^p)_{ad} \\ (\bar{t} - \bar{t}^p)_{ad} \end{Bmatrix}_\ell$$

$$-\frac{iG}{2\pi(1-\nu)}\int_\Gamma N_n^{o_e}\begin{bmatrix}\dfrac{z'}{z^2} & -\dfrac{\bar{z}'}{\bar{z}^2} \\[2ex] \dfrac{\bar{z}'}{z^2}-2\dfrac{\bar{z}z'}{z^3} & -\dfrac{z'}{z^2}\end{bmatrix}\mathrm{d}\xi\begin{Bmatrix}d-d^p \\ \bar{d}-\bar{d}^p\end{Bmatrix}_n \tag{7}$$

which is just a very compact way of writing eqn (2) – thus comprising all subsequent, auxiliary expressions. The displacement and traction parameters at *nodes* n and *loci* ℓ are complex values and their respective conjugate representations, as for displacements at a node n: $d = d_x + id_y$. The stress components are advantageously arranged as shown on the left, so that the cumbersome, real-variable arrays of the former section become the indicated simple expressions of the complex $z = x + iy$ and its conjugate. As developed in Dumont [10], we do not need to resort to the literature on complex-variable elasticity to arrive at the above equation, although the compact stress representation is certainly an inspiration from Muskhelishvili [12].

Besides its simplicity, there are some added advantages in using the latter expression. In fact, as shown in Dumont [10], we no longer need to distinguish between real and complex quasi-singularity poles, which simplifies the code implementation substantially. However, the most important advantage is that, as we see in the latter expression, the singularities we have to deal with are of the type $1/z$, $1/z^2$ and $1/z^3$ in terms of the complex pole $\xi_s = a + ib$, whereas in the former section, as given in the last paragraph, we may have quasi-singularities up to $1/r^6$ of the complex double-pole $\xi_s = a \pm ib$. Apart from that, most of the conceptual and code-implementation developments of Dumont [8] are used in Dumont [10].

The complete developments and code-implementation issues for 2D potential and elasticity problems are given in Dumont [8] and [10] in terms of real and complex variables, and are outlined above rather briefly for the most critical case of evaluating results at internal points. Numerical assessments are shown in the following for a simple problem that is made challenging on purpose, as well as for a second, actually extremely challenging, problem.

4 APPLICATION TO AN INFINITE PLATE WITH A HOLE

Fig. 1 shows on the left the geometric discretization of an infinite plate with a hole of unit radius. The solution for a uniform stress field $\sigma_{xx} = 1$, $\sigma_{yy} = \tau_{xy} = 0$ applied at infinity, according to Sadd [13], for instance, is compactly expressed in polar coordinates as

$$\sigma_{rr} = \frac{1}{2}\left(1-\frac{a^2}{r^2}\right)+\frac{1}{2}\left(1-4\frac{a^2}{r^2}+3\frac{a^4}{r^4}\right)(2\cos^2\theta-1)$$

$$\sigma_{\theta\theta} = \frac{1}{2}\left(1+\frac{a^2}{r^2}\right)-\frac{1}{2}\left(1+3\frac{a^4}{r^4}\right)(2\cos^2\theta-1) \tag{8}$$

$$\tau_{r\theta} = \left(-1-2\frac{a^2}{r^2}+3\frac{a^4}{r^4}\right)\sin\theta\cos\theta$$

The maximum stress value is $\sigma_{\max} = \sigma_{\theta\theta}(r = a, \theta = \pm\pi/2) = 3$, and we also obtain the highest normal stress in the vertical direction $\sigma_{\theta\theta}(r = a, \theta = 0) = -1$. For the sake of reference, we express on the right in Fig. 2 the analytical stress results $\sigma_x, \sigma_y, \tau_{xy}$ of eqn (8) at the 18 internal internal points of the open domain given on the left – and as detailed next – with the indicated lines just connecting points.

As indicated in the figure, we discretize the circular cavity with 10 quartic elements and 40 equally spaced, equidistant nodes from the center – not taking advantage of the double symmetry. The main source of errors in this simulation is related to the fact that the circular surface is modeled with quartic polynomial elements, with non-smooth transition between elements. In fact, we measure for nodes 1, 5, 9, 13, 17, 21, 25, 29, 33 and 37 the same lack of

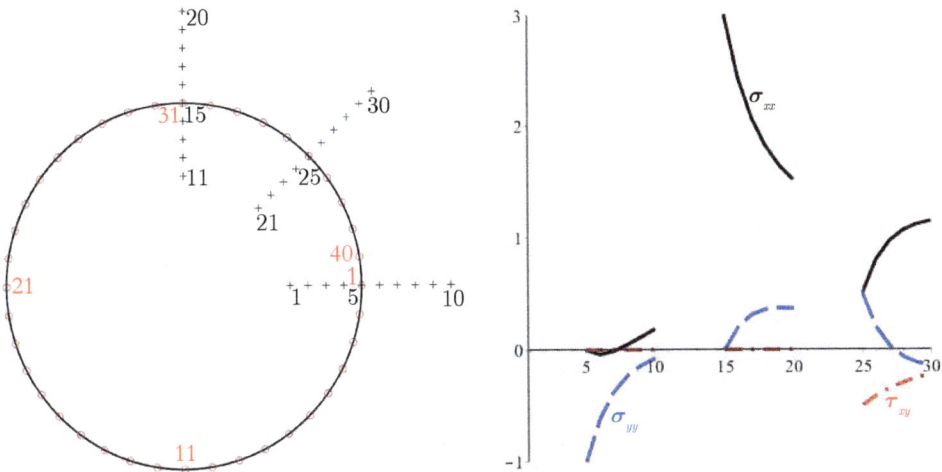

Figure 1: Infinite plate with a circular hole (left) modeled as 10 quartic elements; and analytical stress results evaluated at the indicated 18 internal points.

smoothness, with $(\theta^+ - \theta^-)/2\pi \approx 0.4999900076$, when it should be exactly $1/2$. Then, we should not expect in our numerical evaluations a relative accuracy error smaller than about 0.00001998. When evaluating results, points along a horizontal axis on the left in Fig. 1 that goes through nodes 21 and 1 should present the largest errors, as there is some angularity there. Tangents about nodes 11 and 31 are horizontal and we should expect the smallest errors measured along a vertical from there. Nodes 6, 16, 26 and 36, which are located at angles multiple of $\pi/4$ of inclination, do not have tangents inclined of exact multiples of $\pi/4$, since they are not middle nodes of the quartic elements. Such lacks of polar symmetry are intentional in order to have simulation errors introduced in our model. The figure also indicates three series of 10 points, each, at which stress results are to be evaluated. Points 1, 11 and 21 are $0.6 + 10^{-10}$ distant from the center, with interpolated points up to the respective points 10, 20 and 30, which are at a distance $1.5 + 10^{-10}$ from the center, in such a way that points $1 \ldots 4$, $11 \ldots 14$ and $21 \ldots 24$ are actually internal to the cavity, that is, external in relation to the open domain of interest. We intentionally set points 5, 15 and 25 inside the open domain and just 10^{-10} distant from the respective nodes 1, 31 and 36. Besides the geometry errors, we should expect some round-off errors related to these close points, as assessed next.

In both real- and complex-variable codes we use 8 Gauss–Legendre points per element, thus a total of 80 integration points for the whole problem. We also use 25 digits of precision in the Maple implementation (Maplesoft, a division of Waterloo Maple Inc., Waterloo, Ontario). This is by far more than necessary in the evaluation of matrices $\mathbf{G}$ and $\mathbf{H}$, to be used in eqn (1). In fact, in the precision evaluation $|\mathbf{HW} - \mathbf{W}|$, where $\mathbf{W}$ is a matrix with the three rigid-body displacements that the discretized circular hole may undergo, we obtain the global relative error of just 10^{-18} according to both codes. On the other hand, owing to the very small source-to-boundary distance 10^{-10}, as described above, severe round-off errors may be expected in the evaluation of stress results at points 5, 15 and 25.

Eqn (1) is solved for the displacement difference $(\mathbf{d} - \mathbf{d}^p)$ considering that $\mathbf{t} = \mathbf{0}$ and $\mathbf{t}^p$ corresponds to the proposed far stress field ($\sigma_{xx} = 1, \sigma_{yy} = \tau_{xy} = 0$). Stress results are subsequently evaluated according to eqns (2) and (7). Since the simulated problem corresponds to an applied constant stress field, the solution is considered exact for the proposed geometry, according to Theorem 1 of Dumont [7], within the adopted precision digits and numerical quadrature errors of the regular integrals involved. The distance threshold (relative distance from a source point to an element chord) is set equal to 2, in order to consider the source point sufficiently close as to require analytical corrections, according to Dumont [8]–[10]. This means that the numerical evaluations cannot be more precise than provided by the numerical quadrature – also considering the eventuality of round-off errors.

The graphs of Fig. 2 show the absolute errors – as compared with the analytical values for the circular cavity, according to eqn (8) – of the numerical stress results σ_{xx}, σ_{yy} and τ_{xy} at the 30 external and internal points described above. The dash, blue lines correspond to the real-variable evaluations of Section 3.1, and the solid, black lines correspond to the complex-variable evaluations of Section 3.2. Actual accuracy is assessed for results at the cavity points 1–4, 11–14 and 21–24, which should be zero regardless of boundary geometry approximation: the absolute errors – in general smaller than 10^{-6} – are attributable to the relatively low order of Gauss–Legendre quadrature for the large, high-order quartic elements, which may be improved by resorting to a higher quadrature or maybe by choosing a close/far threshold larger than 2. This is, however, not a useful measure, as there is already a larger geometry error in the piece-wise representation of the circular hole, as assessed above. And there is an information to draw from these assessments: our accuracy threshold is about 10^{-6}, which should be our precision threshold, as well. We see in Fig. 2 that the real- and complex-variable simulations deliver approximately the same accuracy of results, except that real-variable round-off-errors become excessively large at the critical, close point 5, while just reasonably large at points 15 and 25. On the other hand, the largest complex-variable, absolute errors of $(\sigma_{xx}, \sigma_{yy}, \tau_{xy})$ at point 5, close to an angularity and to be compared

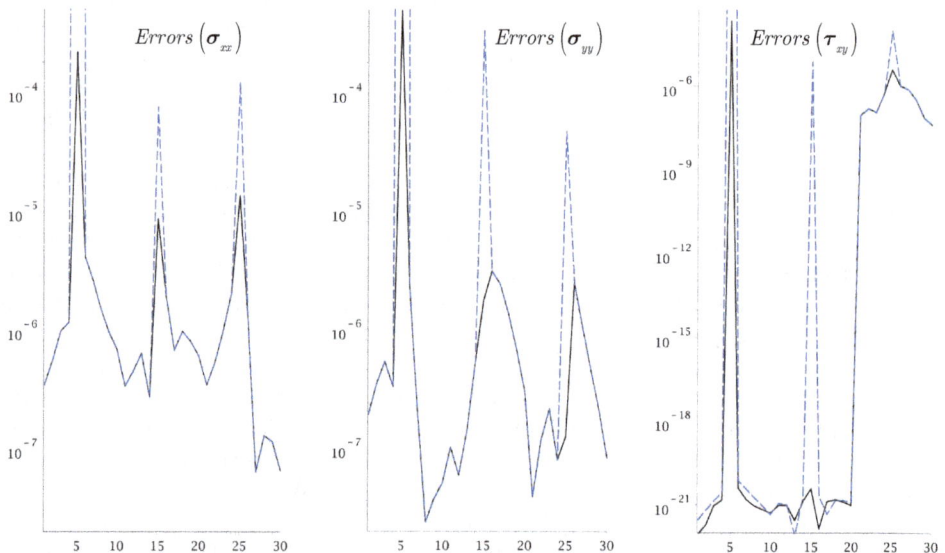

Figure 2: Absolute errors in the numerical evaluation of stress results σ_{xx}, σ_{yy}, τ_{xy} at 30 internal/external points for the infinite plate with a hole of Fig. 1, in terms of complex (solid, black lines) and real variables (dash, blue lines).

with $(0, -1, 0)$, are $(2.19 \times 10^{-4}, 4.92 \times 10^{-4}, 2.08 \times 10^{-4})$. At point 15, close to node 31 of a smooth surface and for stresses to be compared with $(3, 0, 0)$, these errors are $(6.81 \times 10^{-6}, 1.77 \times 10^{-6}, 3.78 \times 10^{-21})$. The errors at close point 25, for stresses to be compared with $(0.5, 0.5, -0.5)$, are $(1.35 \times 10^{-5}, 1.25 \times 10^{-7}, 3.87 \times 10^{-6})$.

5 APPLICATION TO A TOPOLOGICALLY EXTREMELY CHALLENGING PROBLEM

Fig. 3 represents a two-dimensional domain (about 25 units across) with some challenging topological features, to be subjected to a series of elastic fields, as described in Dumont [10], which is on the other hand a development of a simpler numerical model proposed in Dumont [9]. Readers are referred to these papers for the complete description of the problem. It is worth remarking that the cusp at node 1 has an internal angle of about 10^{-8} *rad*, the external angle at node 17 is of about 10^{-13} *rad*, and the strip of material between the cavity and the external boundary is only about 10^{-4} unities wide. This elastic body is subjected to two constant and a series of four linear, quadratic, cubic, quartic and quintic polynomial fields, thus a total of 22 fundamental solutions of the elastostatics problem for homogeneous, isotropic material with shear modulus $G = 80000$ and Poisson's ratio $\nu = 0.2$. The indicated crosses in Fig. 3 are a total of 41 – in part internal and in part external – points at which stress results are to be numerically evaluated for the applied stress fields. Some of these points are very close to the boundary, as described in Dumont [9], [10] and to be assessed next. Most important, we generate between internal point 30 and node 69, which are visually indistinguishable from each other in the figure, a series of 10 points that approach node 69 at geometrically decreasing distances, as indicated in the first row of Table 1, which is a reproduction from Dumont [10]. A similar series of 10 very close points to node 17 is also generated, with the distances indicated in the second row of this table. If we consider all

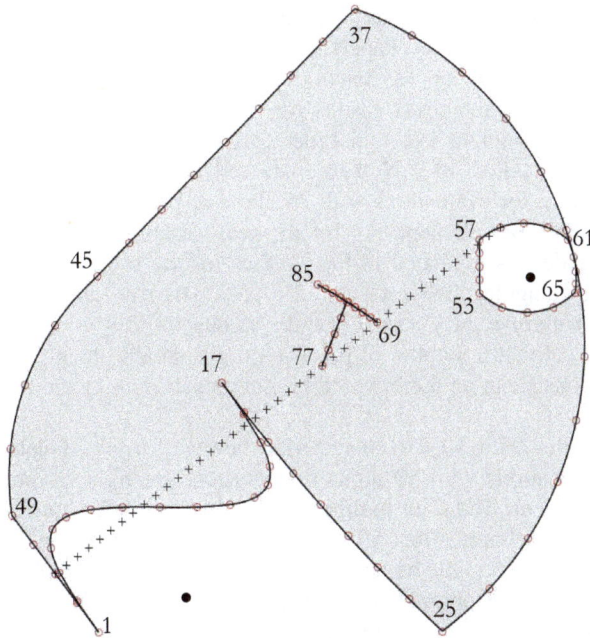

Figure 3: Two-dimensional figure with some challenging topological issues and to be subjected to a series of elastic fields, as reproduced from Dumont [10].

Table 1: Distances from very close points to their reference nodes in a mesh with 92 nodes, Fig. 3, as detailed in Dumont [10], from which this table is reproduced. The internal points 30 and P_0, with coordinates $\approx (16.27636, 17.35048)$ and $(7.19999, 14.00001)$, are close to nodes 69 and 17, respectively. The indicated distances refer to internal points numbered 42–51, on the left, and 52–61.

Internal $30 \rightarrow 69$	Internal $P_0 \rightarrow 17$
5.8947×10^{-17}	1.3120×10^{-18}
1.2378×10^{-15}	2.7552×10^{-17}
2.4816×10^{-14}	5.5235×10^{-16}
4.9639×10^{-13}	1.1048×10^{-14}
9.9280×10^{-12}	2.2097×10^{-13}
1.9856×10^{-10}	4.4194×10^{-12}
3.9712×10^{-9}	8.8388×10^{-11}
7.9424×10^{-8}	1.7677×10^{-9}
1.5884×10^{-6}	3.5355×10^{-8}
3.1769×10^{-5}	7.0710×10^{-7}

geometric data given in m, the smallest distances are actually about one thousandth of a typical proton size! As remarked in Dumont [10]: "This has absolutely no sense in terms of continuum mechanics but the mathematics must work nevertheless."

Dumont [9] deals with a similar problem to the one of Fig. 3 in terms of real variables, which is taken from Dumont [10] for a series of assessments of consistency and accuracy of eqn (1) as well as for displacement (not shown here) and stress results at internal points, using eqns (2) and (7), for different boundary element orders, numbers of Gauss–Legendre quadrature points, and precision digits. Papers [9] and [10] also have complete problem formulation and numerical assessments for potential problems.

We assess in the following the stress results for two numerical models considered in Dumont [10] – but not in Dumont [9] – in order to compare the proposed approaches directly. The first model is called 46_2_25_4 to characterize the use of $n_e = 46$ elements of order $o_e = 2$ (quadratic), for evaluations with 25 digits of precision and $n_g = 4$ Gauss–Legendre quadrature points per element. As for the real-variable formulation, there is a total of $n_e \times o_e = 92$ nodes, as indicated in Fig. 3, thus for the square matrix $\mathbf{H}$ of order $2 \times 92 = 184$ and the rectangular matrix $\mathbf{G}$ with $2 \times 92 = 184$ rows and $2 \times (92 + 46) = 276$ columns. The total number of Gauss–Legendre points is $46 \times 4 = 184$. It is worth repeating from Dumont [9] that a number of precision digits smaller than 20 would render the present problem inconsistent, as there would be interpenetration of the boundary faces adjacent to node 17.

A second model is called 23_4_50_8 to characterize the use of $n_e = 23$ elements of order $o_e = 4$ (quartic), for evaluations with 50 digits of precision and $n_g = 8$ Gauss–Legendre quadrature points per element. Then, all numbers given for the former model apply here, with the only – relevant – difference that 50 precision digits are used, and, of course, not neglecting that quartic elements enable better modeling than double the number of quadratic elements, as already assessed in Dumont [9] and [10].

The indicated matrix dimensions are actually not relevant for the only kind of numerical assessments we are showing, namely, the evaluation of stress results at internal/external points, according to eqns (2) and (7), for $d_n^p = 0$, $t_\ell^p = 0$, and the boundary displacement and traction parameters d_n and t_ℓ evaluated for the 22 polynomial fields of fundamental solutions.

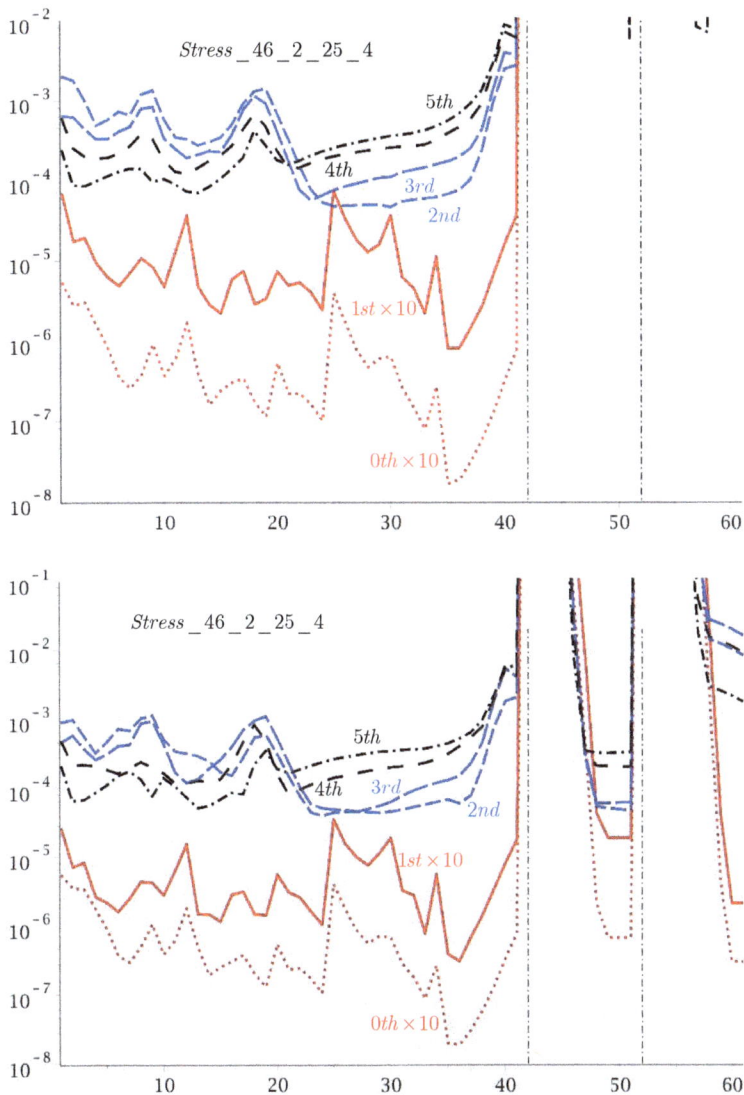

Figure 4: Errors in the evaluation of stress results along 61 internal/external points of Fig. 3, generated for six groups of elasticity test fields for the mesh configuration 46_2_25_4, in terms of real (top) and complex (bottom) variables. The bottom plot is a reproduction from Dumont [10].

Figs 4 and 5 show error results for stress evaluations at the 41 points marked as "+" in Fig. 3 (of which points 2 through 8 and 40 are actually external), as well as for the very close points numbered 42 through 61 referred to in Table 1, as given in the horizontal axes. Vertical dashdot lines are drawn to mark the closest points 42 and 52. Relative errors are given in the vertical log scale for groups of two constant displacements solutions (0th) and sets of four linear, quadratic, cubic, quartic and quintic solutions (1st, 2nd, 3rd, 4th and 5th, respectively). As given in Theorem 1 of Dumont [7] and pointed out in Dumont [9] and [10], the numerical

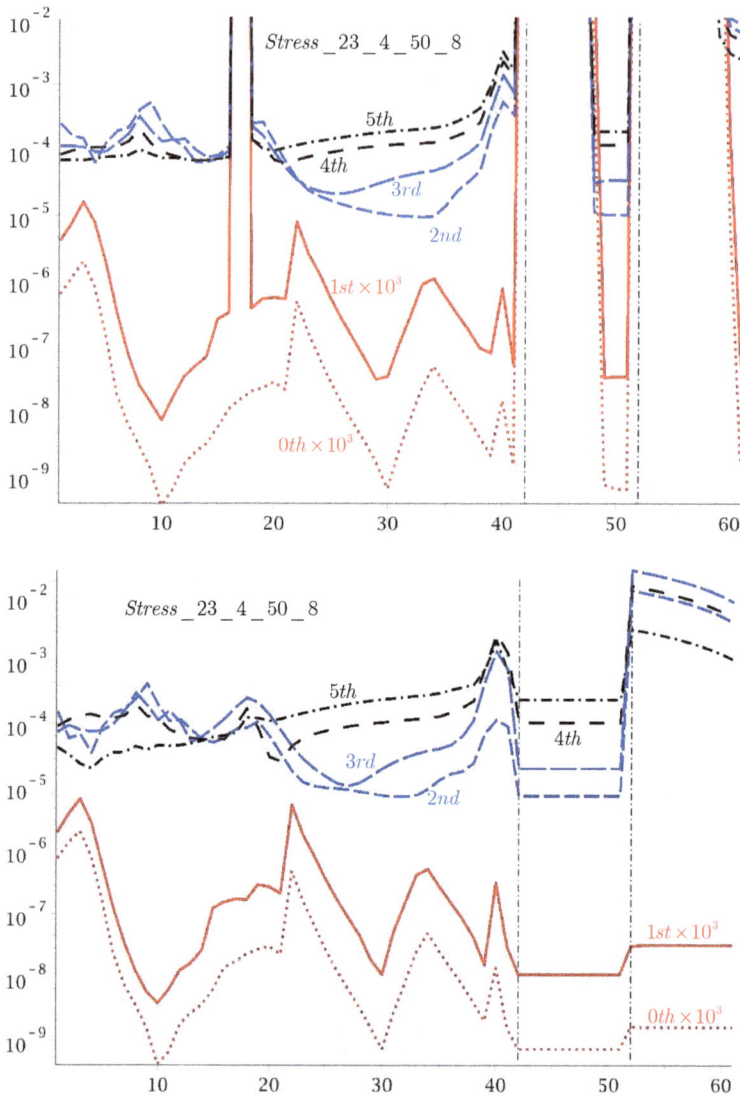

Figure 5: Errors in the evaluation of stress results along 61 internal/external points of Fig. 3, generated for six groups of elasticity test fields for the mesh configuration 23_4_50_8, in terms of real (top) and complex (bottom) variables. The bottom plot is a reproduction from Dumont [10].

evaluations for constant and linear solutions must correspond to analytical solutions of the problem – although within precision digits, number of Gauss–Legendre quadrature points and eventual round-off errors. Then, these solutions give the means of assessing the precision threshold of a numerical simulation. As indicated, the error results for constant and linear displacement fields are multiplied by powers of 10 in order to accommodate a better error resolution in the plots. In both figures the top graphs refer to evaluations in terms of real variables, to be compared with the bottom results in terms of a complex variable.

The real-variable and complex-variable results in each figure are comparable in terms of precision (results for 0th and 1st) and accuracy (for higher displacement fields), except for points that are too close to the boundary, as round-off errors take place and may become completely out of limits. We see in the top plot of Fig. 4 that real-variable results are reliable only for distances larger than about 10^{-7} (which, for unities in m, are smaller than a micrometer), whereas for the complex-variable evaluations we obtain reliable results for subnanometer distances about 10^{-10}. And that for "only" 25 digits of precision!

Fig. 5 presents – for in principle the same computational cost as for the previous evaluations – results that are several orders of magnitude more precise and one order of magnitude more accurate (a not lesser remark is that cost increases for the 50 digits of precision). This is related to the p-refinement from quadratic to quartic elements [9], [10]. Reliable results for the real-variable evaluations are achieved for distances that are orders of magnitude smaller than before, and this is to be credited to the large number of precision digits. On the other hand, results for internal points close to the (badly locally discretized) reentrance tend to be completely out of limits, as well as results for point 40, which is just outside the domain, tend to be more inaccurate, which is also related to the poor local mesh discretization. On the other hand, it is highly remarkable that the complex-variable evaluations shown on the bottom of Fig. 5 lead to reliable results up to the incredibly small distance of 10^{-18}. In fact, we see that the precision assessments for constant and linear displacement fields – and, then, Theorem 1 of Dumont [7] – holds just fine. Observe that here, too, the external point 40 is close to a locally coarse mesh, and consequently with relatively worse results. Moreover, observe that results at points 42 through 51 (piece-wise horizontal plots) are orders of magnitude more accurate than results at points 52 through 61 for the very simple reason that the corresponding local mesh – adjacent to node 17 – is very coarse.

6 CONCLUDING REMARKS

This brief outline and the simple, comparative numerical assessments show that the paradigm-shifting developments proposed in Dumont [7]–[9] may be still improved – in terms of code simplicity, cost and overall robustness – if we resort to a complex-variable formulation [10], as for 2D elasticity. The numerical illustrations refer once more to a highly important convergence Theorem 1 [7]; separately assess precision, accuracy, and liability to round-off errors; show that mesh refinement is required only to improve mechanical simulation, not for mathematical evaluations; and also show that we should not shun extreme cases of topological issues. A not lesser contribution is the illustration of the way three distinct geometric entities must be properly handled: *boundary nodes n* (for displacements), *boundary loci ℓ* (to which tractions are referred), and *domain points s*, at which we collocate the singular sources.

ACKNOWLEDGEMENT

This project was supported by the Brazilian federal agencies CAPES and CNPq, as well as by the state agency FAPERJ.

REFERENCES

[1] Dumont, N.A., The hybrid boundary element method: An alliance between mechanical consistency and simplicity. *Applied Mechanics Reviews*, **42**(11), pp. S54–S63, 1989.
[2] Dumont, N.A., On the efficient numerical evaluation of integrals with complex singularity poles. *Engineering Analysis with Boundary Elements*, **13**, pp. 155–168, 1994.

[3] Dumont, N.A. & Noronha, M., A simple, accurate scheme for the numerical evaluation of integrals with complex singularity poles. *Computational Mechanics*, **22**(1), pp. 42–49, 1998.

[4] Dumont, N.A., An assessment of the spectral properties of the matrix G used in the boundary element methods. *Computational Mechanics*, **22**(1), pp. 32–41, 1998.

[5] Dumont, N.A., The boundary element method revisited. *Boundary Elements and Other Mesh Reduction Methods XXXII*, ed. C.A. Brebbia, WIT Press: Southampton, pp. 227–238, 2010.

[6] Dumont, N.A., The collocation boundary element method revisited: Perfect code for 2D problems. *International Journal of Computational Methods and Experimental Measurements*, **6**(6), pp. 965–975, 2018.

[7] Dumont, N.A., The consistent boundary element method for potential and elasticity: Part I – Formulation and convergence theorem. *EABE – Engineering Analysis with Boundary Element Methods*, **149**, pp. 127–142, 2023.

[8] Dumont, N.A., The consistent boundary element method for potential and elasticity: Part II – Machine-precision numerical evaluations for 2D problems. *EABE – Engineering Analysis with Boundary Element Methods*, **149**, pp. 92–111, 2023.

[9] Dumont, N.A., The consistent boundary element method for potential and elasticity: Part III – Topologically challenging numerical assessments for 2D problems. *Engineering Analysis with Boundary Element Methods*, **151**, pp. 548–564, 2023.

[10] Dumont, N.A., Complex-variable, high-precision formulation of the consistent boundary element method for 2D potential and elasticity problems. *Engineering Analysis with Boundary Element Methods*, **152**, pp. 552–574, 2023.

[11] Brebbia, C.A., Telles, J.C.F. & Wrobel, L.C., *Boundary Element Techniques*, Springer-Verlag, 1984.

[12] Muskhelishvili, N.I., *Some Basic Problems of the Mathematical Theory of Elasticity*, Noordhoff International Publishing: Leyden, 1957.

[13] Sadd, M.H., *Elasticity, Theory, Applications, and Numerics*, Elsevier, 2005.

SECTION 4
COMPUTATIONAL METHODS

MODELLING OF FIRE RESISTANCE AND MECHANICAL PERFORMANCE OF GLASS FAÇADES

KIM MEOW LIEW[1,2], WEIKANG SUN[1], BINBIN YIN[1,2],
JINHUA SUN[3] & VENKATESH KUMAR R. KODUR[4]
[1]Department of Architecture and Civil Engineering, City University of Hong Kong, China
[2]Centre for Nature-Inspired Engineering, City University of Hong Kong, China
[3]Thermal Science and Energy Engineering, The University of Science and Technology of China, China
[4]Department of Civil and Environmental Engineering, Michigan State University, USA

ABSTRACT

Glass façades are widely used in modern high-rise buildings because of their aesthetic merits and environmentally friendly characteristics. However, collapsing glass in an accidental fire can lead to huge casualties and economic losses. Here, we develop a meshfree computational framework based on peridynamics to investigate the fire resistance and mechanical performance of glass façades. The proposed model is well verified and applied to explore the damage and failure mechanisms of glass façades. The effects of the temperature distribution on the thermomechanical fracture behaviour of glazing are also explored. Results show that cracks tend to initiate at fixed supporting points such as bolts and their propagation paths are greatly influenced by the temperature distributions. Both temperature gradient and local boundary conditions can play a significant role in the thermomechanical cracking of glass façades. This work provides important insights on cracking mechanisms of glass façades during a fire and provide the building and construction companies with new recommendations and design guidelines for improved fire safety.
Keywords: meshless, modelling, fire resistance, mechanical performance, glass façade.

1 INTRODUCTION

Research interest in glazing protection from fire has been growing with the wide application of glass façades in modern high-rise buildings. Glass façade breakage during a fire induces more fresh air to enter the structure, causing catastrophic consequences [1]. To avoid such hazardous scenarios, understanding and predicting the thermomechanical behaviours of glass façades is crucial. Various experimental setups have been adopted to study the heat transfer and breakage processes of different types of glass [2], [3] under diverse fixation conditions [4], [5] during a fire event. However, full-scale experiments are expensive and require many precautions, particularly for the full-field measurements of temperature and displacement during a fire incident. Therefore, an effective and accurate computational framework is urgently demanded to simulate and predict the thermomechanical behaviours of glass façades.

Several numerical methods have been applied to study the heat transfer mechanisms, such as the finite element method (FEM) [6] and the finite volume method (FVM) [7]. However, these grid-based methods face huge challenges when discontinuity (e.g., crack) occurs in the problem domain. Phase field method has shown its calibre in modelling crack propagation in materials, whereas the correct modelling of multiple cracks relies on local mesh update and refinement [8]. Without mesh issues, meshfree methods have been developing fast in the past decades [9]–[13]. As a nonlocal model, peridynamics (PD) is increasingly used to simulate fracture initiation and propagation in both homogeneous and heterogeneous materials [14]–[17] under various loading conditions [18], [19]. However, few works were reported to apply peridynamics in modelling fire-induced structural failures, particularly, the crack propagation in glass façades during fire events.

WIT Transactions on Engineering Sciences, Vol 135, © 2023 WIT Press
www.witpress.com, ISSN 1743-3533 (on-line)
doi:10.2495/BE460071

To fill this gap, we propose a three-dimensional peridynamic framework to model the thermomechanical behaviours of glass façades. In our PD method, the governing equations for the mechanical problem, heat transfer problem, and their coupled problem are formulated in integral forms, which naturally incorporate the discontinuity scenarios. Two test cases are presented to demonstrate the effectiveness of the proposed method for modelling thermomechanical deformation and cracking in glass façade.

2 COMPUTATIONAL FRAMEWORKS

2.1 Peridynamic framework

In PD theory, two particles can have a non-local interaction with each other from a finite distance. The major variables in PD theory are shown in Fig. 1.

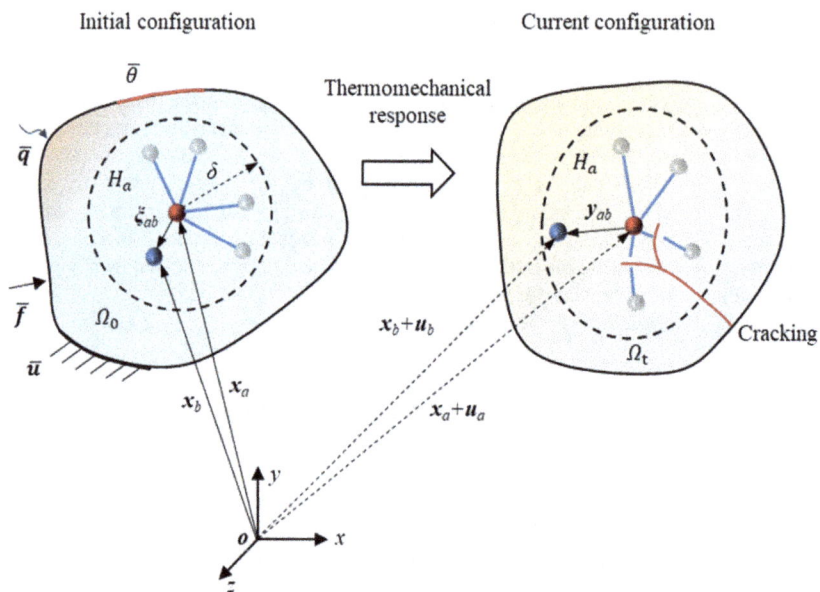

Figure 1: Illustration of variables used in PD framework.

As displayed in Fig. 1, it is assumed in PD that for any given particle (or material point), there is a positive number δ, called the horizon, such that

$$|\boldsymbol{\xi}_{ab}| > \delta \Rightarrow \boldsymbol{f}(\boldsymbol{\xi}_{ab}, \boldsymbol{\eta}_{ab}) = \boldsymbol{0}, \text{ for } \forall \boldsymbol{\xi}_{ab}, \tag{1}$$

where $\boldsymbol{f}$ is the pairwise bond force per unit volume, $\boldsymbol{\xi}_{ab}$ and $\boldsymbol{\eta}_{ab}$ are the relative position and displacement vectors in the reference configuration, which can be written as

$$\boldsymbol{\xi}_{ab} = \boldsymbol{x}_b - \boldsymbol{x}_a, \qquad \boldsymbol{\eta}_{ab} = \boldsymbol{u}_b - \boldsymbol{u}_a. \tag{2}$$

By applying the Newton's second law, the equation of motion in PD can be written as

$$\rho_a \ddot{\boldsymbol{u}}_a = \int_{\mathcal{H}_a} \boldsymbol{f}(\boldsymbol{\xi}_{ab}, \boldsymbol{\eta}_{ab}) \, dV_b + \boldsymbol{b}_a \tag{3}$$

where ρ_a and $\ddot{u}_a$ are the mass density and acceleration of particle a, dV_b is the volume of particle b in the family of particle a (denoted as $\mathcal{H}_a$), and b_a is the external body force applying on particle a.

For an isotropic linear elastic material,

$$f(\xi_{ab}, \eta_{ab}) = c s_{ab} \cdot [1 - d(\xi_{ab}, \eta_{ab})] \cdot \widehat{y_{ab}}, \tag{4}$$

where the unit deformed bond vector

$$\widehat{y_{ab}} = \frac{y_{ab}}{y_{ab}}, y_{ab} = \xi_{ab} + \eta_{ab}, y_{ab} = |y_{ab}| = \lambda_{ab} |\xi_{ab}|, \tag{5}$$

and the bond strain and bond stretch are

$$s_{ab} = (\lambda_{ab} - 1), \lambda_{ab} = \frac{|y_{ab}|}{|\xi_{ab}|}. \tag{6}$$

In eqn (4), c is micro-modulus (or bond stiffness) and $d(\xi_{ab}, \eta_{ab})$ is the bond damage (i.e., $d = 0$ for intact and 1 for broken). The material parameter c can be calibrated with classical theory on the basis of energy equivalence under the small deformation assumption [20], [21] as

$$c = \frac{14E}{\pi \delta^4} \left[1 - \left(\frac{|\xi_{ab}|}{\delta} \right)^2 \right]^2, \tag{7}$$

for 3D cases, where E is the Young's modulus, v is the Poisson's ratio and h is the thickness of the body in a 2D problem. The bond damage $d(\xi_{ab}, \eta_{ab})$ is defined as

$$d(\xi_{ab}, \eta_{ab}) = \begin{cases} 0 & if \ s_{ab} < s_c, \\ 1 & if \ s_{ab} \geq s_c, \end{cases} \tag{8}$$

where s_c is the critical bond stretch, which can be calibrated using linear fracture mechanics parameter (i.e., the critical energy release rate G_c) [20], [21] or the tensile strength σ_b (used in this work) or their minimum value as

$$s_c = \min \left(\frac{\sigma_b}{E}, \sqrt{\frac{1024 \pi G_c}{7(120\pi - 133)E\delta}} \right). \tag{9}$$

The damage of a particle a can be further defined as

$$D_a = \frac{\sum_{b=1}^{N_a} d(\xi_{ab}, \eta_{ab})}{N_a}, \tag{10}$$

where $d(\xi_{ab}, \eta_{ab})$ is the bond damage defined in eqn (8) and N_a is the number of total bonds in the family of particle a.

2.2 PD for heat transfer

To derive the PD version of heat transfer equation, we first recall the energy balance equation for a purely heat transfer process

$$\int_\Omega \rho c_v \frac{\partial \theta}{\partial t} dv = - \int_{\partial \Omega} q \cdot n \, da + \int_\Omega q_s \, dv, \tag{11}$$

where ρ is the mass density, c_v is the specific heat capacity, q is the heat flux, n is the unit surface normal vector and q_s is the source term for heat generation. Applying the divergence theorem and the Fourier's law for heat conduction $\boldsymbol{q} = -\kappa \boldsymbol{\nabla}\theta$ (κ is the heat conduction coefficient), we have the local form of heat transfer equation

$$\rho c_v \frac{\partial \theta}{\partial t} = \kappa \Delta \theta + q_s. \tag{12}$$

In PD theory, we have

$$\rho_a c_v \frac{\partial \theta_a}{\partial t} = \int_{\mathcal{H}_a} K(\boldsymbol{x}_a, \boldsymbol{x}_b) \frac{\theta_b - \theta_a}{\|x_b - x_a\|^2} dV_b + h_s,$$

$$\tag{13}$$

where $K(\boldsymbol{x}_a, \boldsymbol{x}_b)$ is the heat conduction coefficient of PD thermal bond between particles a and b, and h_s is the source term of heat generation per unit mass. By drawing an equivalence between PD and classical heat flux, $K(\boldsymbol{x}_a, \boldsymbol{x}_b)$ can be calibrated and expressed with classical parameter k. For 3D cases, $k_0 = 9\kappa/(2\pi\delta^3)$ and $k_1 = 18\kappa/(\pi\delta^3)$.

2.3 Initial and boundary conditions

In terms of heat transfer, the initial condition can be implemented directly on each PD particles as

$$\theta_a(\boldsymbol{x}, t)|_{t=0} = \theta_0(\boldsymbol{x}). \tag{14}$$

A uniform Dirichlet boundary condition will also be used in this work, which can be written as

$$\theta(\boldsymbol{x}, t)|_{x \in \Omega_d} = \theta^*. \tag{15}$$

where θ^* is again the prescribed temperature at the Dirichlet boundary $\boldsymbol{x} = \boldsymbol{x}^*$ and Ω_d is the Dirichlet boundary layer generated by mirroring PD particles symmetric to $\boldsymbol{x} = \boldsymbol{x}^*$.

Whereas the Neumann boundary conditions will be modified before applied to boundary PD particles in Neumann boundary layers Ω_n. The principle of modification is to assume that the heat flux $\boldsymbol{q}$ is uniform through the thickness Δ of boundary layer and then the heat generation per unit volume can be derived from the heat flux according to the energy balance law

$$\int_{\Omega_n} \rho_a c_v \frac{\partial \theta_a}{\partial t} dV_a = - \int_{\partial\Omega_n} \boldsymbol{q} \cdot \boldsymbol{n} \, dS_a, \tag{16}$$

as

$$\rho_a c_v \frac{\partial \theta_a(x,t)}{\partial t}\Big|_{x \in \Omega_n} = -\frac{q \cdot n S_a}{V_a} = -\frac{q \cdot n}{\Delta}. \tag{17}$$

Particularly, for convective boundary conditions

$$\rho_a c_v \frac{\partial \theta_a(x,t)}{\partial t}\Big|_{x \in \Omega_n} = \frac{h}{\Delta}(\theta_\infty - \theta_a), \tag{18}$$

where θ_∞ is the temperature of surrounding medium, and h is the convective heat transfer coefficient.

Furthermore, for radiation boundary conditions

$$\rho_a c_v \frac{\partial \theta_a(x,t)}{\partial t}\Big|_{x \in \Omega_n} = \frac{\varepsilon\sigma}{\Delta}(\theta_s^4 - \theta_a^4), \tag{19}$$

where θ_s is the temperature of the surface surrounding the body, ε is emissivity of the boundary surface and σ is the Stefan–Boltzmann constant.

On the other hand, the mechanical initial and boundary conditions are rather straight forward (applying displacement, velocity and body force on boundary particles) [16].

2.4 Thermomechanical coupling

Thermomechanical coupling has two aspects, i.e., thermal strain (from temperature field to strain field) and mechanical heating (from strain rate field to temperature field). In this work, only the former one is considered. By inserting the thermal strain into eqn (4), the governing equation can be rewritten as

$$\rho_a \ddot{\pmb{u}}_a = \int_{\mathcal{H}_a} c(s_{ab} - \alpha\theta_{avg}) \cdot [1 - d(\pmb{\xi}_{ab}, \pmb{\eta}_{ab})] \cdot \widehat{\pmb{y}_{ab}} \, dV_b + \pmb{b}_a, \tag{20}$$

where α is the thermal expansion coefficient

$$\theta_{avg} = \frac{\theta_a + \theta_b}{2}, \tag{21}$$

and other variables are given in eqns (5)–(8).

3 RESULTS AND DISCUSSION

3.1 Verification test example

This test case is aimed to verify the PD method for modelling thermomechanical deformation.

As shown in Fig. 2, the side length of the square is $L = 1$ cm. The initial temperature is $\theta_0 = 0°C$. The lateral displacement of left and right sides of the square are constrained. The vertical motion of the bottom side of the square is also constrained. A thermal shock of $\theta_1 = 1°C$ is applied to the upper side of the square. The Young's modulus of the square is 1 GPa, Poisson's ratio is 0.33, thermal expansion coefficient is 0.02/K, thermal conductivity is 1 W/(m·K) and specific heat capacity is 1 J/(kg·K).

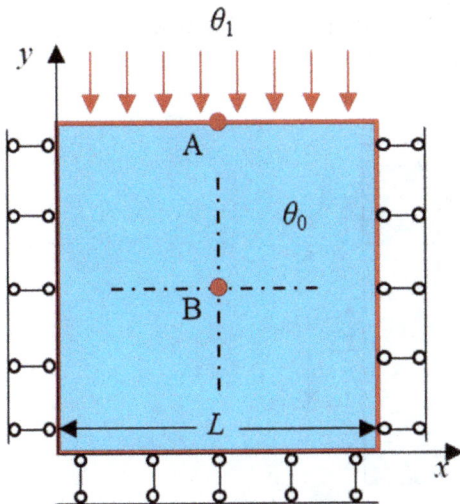

Initial conditions: for $t = 0$,

$$\theta(x, y) = \theta_0.$$

Boundary conditions: for $\forall\, t > 0$,

$$\theta|_{y=L} = \theta_1,$$

$$u_y|_{y=0} = u_x|_{x=0} = u_x|_{x=L} = 0.$$

Figure 2: Test setup of the square plate under thermal shock.

The displacement history at point A and the temperature history at point B marked in Fig. 2 are given in Fig. 3. The temperature increase is nonlinear over time. With the increase of spatial resolution, PD results show greater agreement with the analytical results presented in Chen et al. [22], confirming the convergence and accuracy of the proposed PD method.

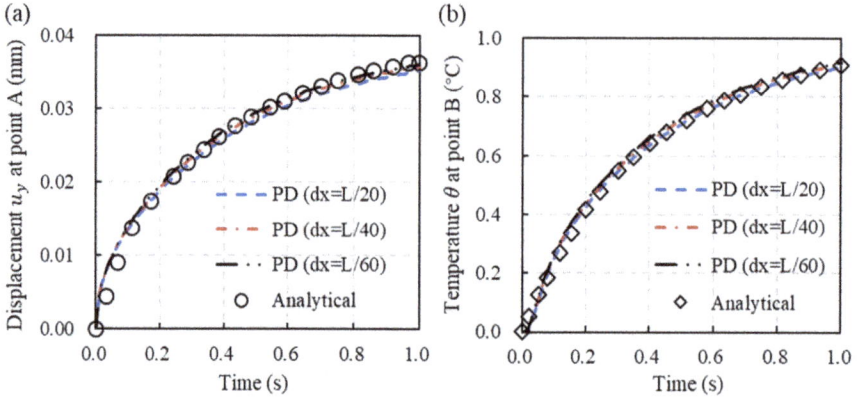

Figure 3:   Comparison of PD and analytical results of displacement at point A and temperature history at point B.

3.2 Effects of temperature distribution on the fracture pattern of glazing

To explore the effect of temperature distribution and fixture forms, a square single glazing with side length of 1.2 m, thickness of 6 mm and four-point fixture (by bolts) was studied. As shown in Fig. 4, the distance between two bolts D is 0.2 m and three levels of the width of high temperature zone $H = L/4$, $L/3$ and $L/2$ are adopted. The Young's modulus of the glazing is 67 GPa, Poisson's ratio is 0.25, tensile strength is 40 MPa, and thermal expansion coefficient is 8.5×10^{-6}/K. Temperature increasing rates are 1.2 and 0.5 K/s for high and low temperature zones respectively. The horizon radius is 3.015 times the initial particle spacing. A total number of 121,203 particles are used in this example.

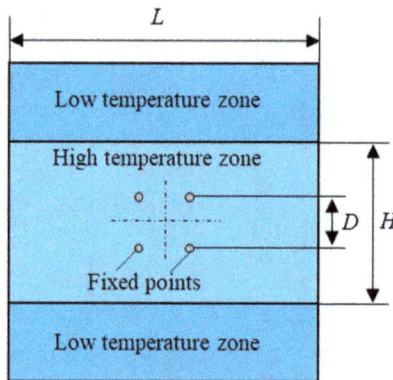

Figure 4:   Test setup of the square single glazing. (a) Temperature distribution; and (b) Geometry.

The crack patterns of the glazing with different temperature distributions are presented in Fig. 5. As can be observed, crack initiates at the fixed points for all the three test cases. However, once initiated, the crack propagation paths are remarkably different for different widths of the high-temperature zone. For the case of $H = L/4$, initiated cracks first grow into two parallel cracks in the direction vertical to the two temperature–transition interfaces. After these cracks meet the two interfaces, they propagate along the interfaces until they fully penetrate the glazing. For the case of $H = L/3$, the crack pattern is quite similar despite that the cracks form curved paths instead straight ones before meeting the temperature–transition interfaces. For the case of $H = L/2$, the crack pattern is much more complex. Cracks not only form bended paths before meeting the temperature–transition interfaces they also merge into a straight path parallel to the temperature–transition interfaces. In addition, the cracks at the temperature–transition interfaces bifurcate before penetrating the glazing. These results imply that both the mechanical factors (e.g., point supports or frame supports) and the thermal factors (e.g., the temperature distribution and gradient) play significant roles in the cracking mechanisms of glass façades.

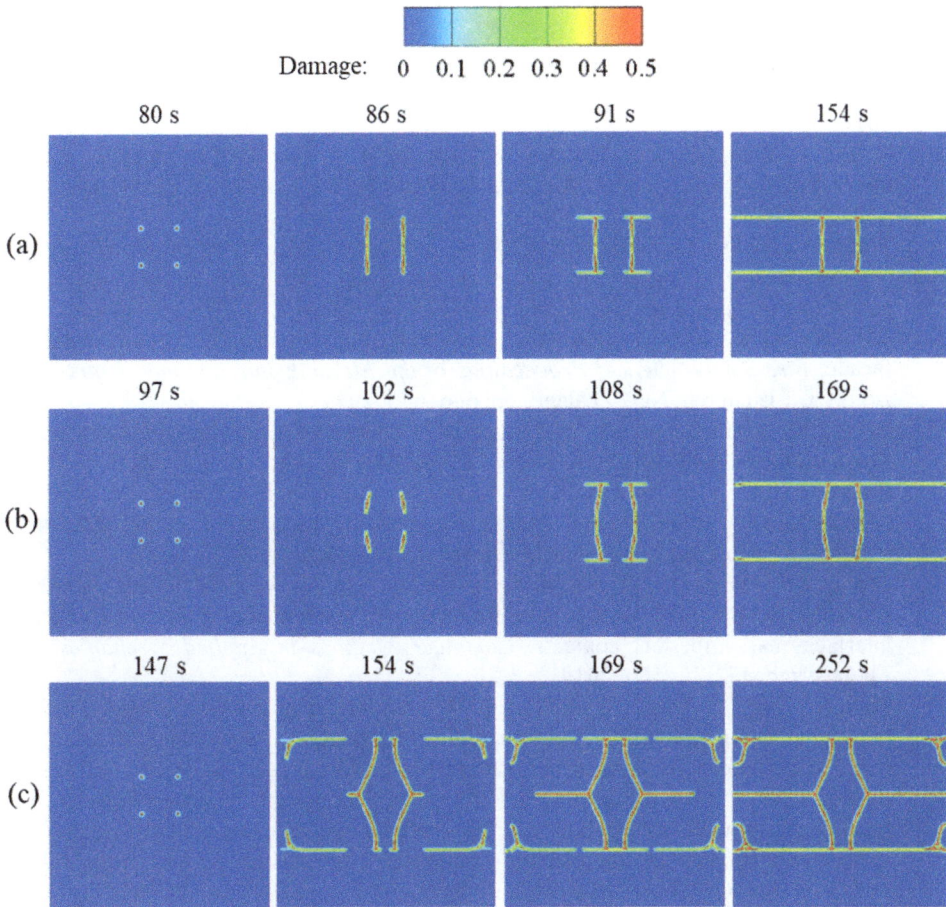

Figure 5:   Crack evolution processes of the square glazing of different temperature distribution with (a) $H = L/4$; (b) $H = L/3$; and (c) $H = L/2$.

4 CONCLUSIONS

In this paper, we present a peridynamic framework to simulate the thermomechanical behaviours of glass façades exposed to fire. The computational effectiveness and accuracy of our PD method was verified by the thermomechanical deformation of a square plate. The proposed PD method was applied to explore the effect of temperature distribution on the crack patterns of glazing. Results indicate that both the mechanical factors such as point supports and the thermal factors like the temperature gradient play significant roles in the cracking of glass façades. To predict the thermomechanical and fracture behaviour of glass façades more accurately, the realistic boundary conditions at supporting frame or fixed points should be carefully considered and the temperature distribution in the glazing should be accurately measured or modelled.

ACKNOWLEDGEMENT

The authors gratefully acknowledge the supports provided by the Research Grants Council of the Hong Kong Special Administrative Region, China (Project No. 9043135, CityU 11202721).

REFERENCES

[1]   Bedon, C., Structural glass systems under fire: overview of design issues: Experimental research and developments. *Advances in Civil Engineering*, 2120570, 2017.

[2]   Wang, Y., Wang, Q., Sun, J., He, L. & Liew, K.M., Influence of fire location on the thermal performance of glass façades. *Applied Thermal Engineering*, **106**, pp. 438–442, 2016.

[3]   Wang, Y., Wang, Q., Wen, J.X., Sun, J. & Liew, K.M., Investigation of thermal breakage and heat transfer in single, insulated and laminated glazing under fire conditions. *Applied Thermal Engineering*, **125**, pp. 662–672, 2017.

[4]   Wang, Y. et al., Experimental study on thermal breakage of four-point fixed glass façade. *Fire Safety Science: Proceedings of the Eleventh International Symposium, IAFSS*, Christchurch, New Zealand, pp. 666–676, 2014.

[5]   Wang, Y. et al., Numerical study on fire response of glass façades in different installation forms. *Construction and Building Materials*, **61**, pp. 172–180, 2014.

[6]   Bakalakos, S., Kalogeris, I. & Papadopoulos, V., An extended finite element method formulation for modeling multi-phase boundary interactions in steady state heat conduction problems. *Composite Structures*, **258**, 113202, 2021.

[7]   Costa, R., Nobrega, J.M., Clain, S. & Machado, G.J., Very high-order accurate polygonal mesh finite volume scheme for conjugate heat transfer problems with curved interfaces and imperfect contacts. *Computer Methods in Applied Mechanics and Engineering*, **357**, 112560, 2019.

[8]   Tian, F., Tang, X., Xu, T., Yang, J. & Li, L., A hybrid adaptive finite element phase-field method for quasi-static and dynamic brittle fracture. *International Journal for Numerical Methods in Engineering*, **120**(9), pp. 1108–1125, 2019.

[9]   Zhang, L.W., Lei, Z.X. & Liew, K.M., Computation of vibration solution for functionally graded carbon nanotube-reinforced composite thick plates resting on elastic foundations using the element-free IMLS-Ritz method. *Applied Mathematics and Computation*, **256**, pp. 488–504, 2015.

[10] Zhang, L.W. & Liew, K.M., Large deflection analysis of FG-CNT reinforced composite skew plates resting on Pasternak foundations using an element-free approach. *Composite Structures*, **132**, pp. 974–983, 2015.

[11] Zhang, Z., Hao, S.Y., Liew, K.M. & Cheng, Y.M., The improved element-free Galerkin method for two-dimensional elastodynamics problems. *Engineering Analysis with Boundary Elements*, **37**(12), pp. 1576–1584, 2013.

[12] Liew, K.M., Huang, Y.Q. & Reddy, J.N., Moving least squares differential quadrature method and its application to the analysis of shear deformable plates. *International Journal for Numerical Methods in Engineering*, **56**(15), pp. 2331–2351, 2003.

[13] Zhang, Z., Li, D.M., Cheng, Y.M. & Liew, K.M., The improved element-free Galerkin method for three-dimensional wave equation. *Acta Mechanica Sinica*, **28**, pp. 808–818, 2012.

[14] Silling, S.A., Reformulation of elasticity theory for discontinuities and long-range forces. *Journal of the Mechanics and Physics of Solids*, **48**(1), pp. 175–209, 2000.

[15] Silling, S.A., Epton, M., Weckner, O., Xu, J. & Askari, E., Peridynamic states and constitutive modeling. *Journal of Elasticity*, **88**, pp. 151–184, 2007.

[16] Oterkus, S., Madenci, E. & Agwai, A., Fully coupled peridynamic thermomechanics. *Journal of the Mechanics and Physics of Solids*, **64**, pp. 1–23, 2014.

[17] Kilic, B., Peridynamic theory for progressive failure prediction in homogeneous and heterogeneous materials. The University of Arizona, 2008.

[18] Sun, W.K., Zhang, L.W. & Liew, K.M., A coupled SPH-PD model for fluid–structure interaction in an irregular channel flow considering the structural failure. *Computer Methods in Applied Mechanics and Engineering*, **401**, 115573, 2022.

[19] Yin, B.B., Sun, W.K., Zhang, Y. & Liew, K.M., Modeling via peridynamics for large deformation and progressive fracture of hyperelastic materials. *Computer Methods in Applied Mechanics and Engineering*, **403**, 115739, 2023.

[20] Huang, D., Lu, G. & Qiao, P., An improved peridynamic approach for quasi-static elastic deformation and brittle fracture analysis. *International Journal of Mechanical Sciences*, **94**, pp. 111–122, 2015.

[21] Huang, D., Lu, G., Wang, C. & Qiao, P., An extended peridynamic approach for deformation and fracture analysis. *Engineering Fracture Mechanics*, **141**, pp. 196–211, 2015.

[22] Chen, W., Gu, X., Zhang, Q. & Xia, X., A refined thermo-mechanical fully coupled peridynamics with application to concrete cracking. *Engineering Fracture Mechanics*, **242**, 107463, 2021.

COMPARISON OF THE BOUNDARY ELEMENT METHOD AND THE METHOD OF FUNDAMENTAL SOLUTIONS FOR ANALYSIS OF POTENTIAL AND ELASTICITY PROBLEMS IN CONVEX AND CONCAVE DOMAINS

SORAYA ZENHARI[1], MOHAMMAD RAHIM HEMATIYAN[2], AMIR KHOSRAVIFARD[2],
MOHAMMAD REZA FEIZI[3] & HANS-CHRISTIAN MÖHRING[1]
[1]Institut für Werkzeugmaschinen (IW), Universität Stuttgart, Germany
[2]Mechanical Engineering Department, Shiraz University, Iran
[3]Civil and Environmental Engineering Department, Shiraz University, Iran

ABSTRACT
The boundary element method (BEM) and the method of fundamental solutions (MFS) are well-known fundamental solution-based methods for solving a variety of problems. Both methods are boundary-type techniques and can provide accurate results. In comparison to the finite element method (FEM), which is a domain-type method, the BEM and the MFS need less manual effort to solve a problem. The aim of this study is to compare the accuracy and reliability of the BEM and the MFS. This comparison is made for 2D potential and elasticity problems with different boundary and loading conditions. In the comparisons, both convex and concave domains are considered. Both linear and quadratic elements are employed for boundary element analysis of the examples. The discretization of the problem domain in the BEM, i.e., converting the boundary of the problem into boundary elements is relatively simple; however, in the MFS, obtaining appropriate locations of collocation and source points need more attention to obtain reliable solutions. The results obtained from the presented examples show that both methods lead to accurate solutions for convex domains, whereas the BEM is more suitable than the MFS for concave domains.
Keywords: boundary element method, method of fundamental solutions, elasticity, potential problem, convex domain, concave domain.

1 INTRODUCTION
Numerical methods developed for solving partial differential equations can be divided into two basic categories namely domain-type and boundary-free methods. Among domain-type methods, one can mention the finite element method (FEM), the finite difference method, and the finite volume method. The main advantages of boundary-type methods over domain-type methods are their fewer amounts of input data and their potential to tackle with moving boundaries, moving loads, and infinite domains. Over last three decades, engineers and scientists have become more interested in mesh reduction methods. Computational methods such as the boundary element method (BEM) and the method of fundamental solutions (MFS) are among these techniques and both are capable of providing excellent results for solving different problems in physics and engineering [1], [2]. The two methods are based on the knowledge of a fundamental solution of the problem. However, unlike the BEM, the MFS is an integration-free method and has some attractiveness for solving some problems [2], [3]. The MFS reduces the computation time significantly in comparison with the BEM as it does not require boundary discretization and computation of singular or regular integrals over the boundary [4], [5] and it can solve certain inverse problems without iteration [6]. Primarily research on the BEM for potential problems dates back to 1903 when Fredholm established his investigation on potential problems based on a discretization technique. The direct boundary integral equation for potential problems was first proposed by Jaswon [7] and Symm [8]. The BEM for solving potential problems in axisymmetric domains was

successfully established by Rizzo and Shippy [9]. Earlier attempts to use the BEM for elasticity problems was made by Muskhelishvili et al. [10], Kupradze [11]. After that, the BEM was employed for analysis of a wide range of problems.

The idea of the MFS as a computational tool for solving various partial differential equations dates back to 1960s where Kupradze and Aleksidze [12] introduced this method for boundary value problems and later Mathon and Johnston [13] improved this method for numerical implementation. Karageorghis and Fairweather [14] employed the MFS for axisymmetric potential problems for the first time. Later, it was successfully applied to a variety of potential and elastic problems, see for example Fam and Rashed [15], Young et al. [16], Marin et al. [17], Karageorghis et al. [18], Fan and Li [19], Mohammadi et al. [20]. Suitable arrangement of source points and collocation points in the MFS has been always disputable and under discussion among the researchers. An arrangement of collocation points in the MFS can lead to an accurate solution while another arrangement with a little change may lead to inaccurate results [21]. In the MFS, the location of the pseudo boundary can be fixed and selected before solving the problem, or it may be determined by an optimization process. In former case, a pseudo boundary similar to the boundary of the problem is usually considered [22]–[24]. In the latter case, the optimal pseudo boundary as a circle and a sphere in 2D and 3D problems is found, respectively [25]–[27]. Optimal pseudo boundaries can provide more accurate solutions [28]–[30]; however, it should be mentioned that optimization algorithms can be relatively time consuming [27] and disappear the simplicity and attractiveness off the MFS. Recently, Hematiyan et al. have recommended some remarks for properly determination of the position of source points in the MFS for potential [3] and elasticity [31] problems. In previous investigations, the BEM and the MFS have been compared in some aspects. Tadeu et al. [32] investigated the efficiency of the BEM, the MFS and radial basis function (RBF) method in wave propagation problem in an elastic domain. The MFS was found to provide results with more accuracy. Salgado-Ibarra [33] compared the accuracy of the MFS and the BEM for an elliptical domain considering different boundary conditions. A circle was considered as the pseudo boundary for solving the Laplace equation and the Saint Venant's torsion problem using the MFS. They observed that the MFS was less time-consuming and had much easier implementation. Alves et al. [34] applied the BEM and the MFS to 2D Laplace equation. The accuracy of these methods was studied for several problems with discontinuity of boundary conditions. They observed that the BEM with double layer potential has a better performance in comparison with the MFS for problems with discontinuous boundary conditions. Considering different computational fields like magnetic and electric problems the scientist obtained similar observations, which acknowledged the superiority of the MFS over the BEM for some cases [35], [36]. Similar to the BEM, the MFS technique can be particularly useful for the analysis of acoustic problems, Godinho et al. [37] and Godinho and Soares [38] while the formulation of the MFS is simpler than the BEM. Dyhoum et al. [39] investigated several EIT (electrical impedance tomography) problems and found the BEM more convergent and stable in comparison with the MFS, which imposed some restriction on the arrangement of source points. Liravi et al. [40] analysed elastodynamics behaviour of solid structure, e.g., cylinder and a thin circular shell located in the soil. A new control technique was implemented to discover the optimal distance between collocation points. According to this investigation, the MFS showed a greater sensitivity to the parameters than the BEM.

In this study, the MFS and the direct BEM for solving potential and elasticity problems in 2D convex and concave domains are compared. By performing numerical studies, the advantages and disadvantages of the two methods for convex and concave domains with simple and complicated boundary conditions are highlighted. To the authors' best knowledge, it is the first time that the ability of the MFS and the BEM for different shapes of domain are compared.

2 THE MFS AND THE BEM FOR POTENTIAL AND ELASTOSTATIC PROBLEMS

2.1 The MFS for potential problems

In the MFS for 2D problems, some collocation points must be considered on the physical boundary of the problem in order to satisfy the boundary conditions, followed by source points on the pseudo boundary, which is defined around the physical boundary. Any source point placed on the pseudo boundary corresponds to a fundamental solution, which provides a solution. Due to linearity of the Laplace and elasticity equations, the summation of the fundamental solutions with arbitrary coefficients also satisfies the equations. The coefficients (intensities of sources) can be determined by satisfying the boundary conditions of the problem at the collocation points. This procedure generates a system of equations, solution of which provides the unknown coefficients. The number of unknowns and the number of equations is in accordance with the number of source points and collocation points, respectively. The solution for an arbitrary internal or boundary point can be found by taking into account the effects of all sources on the pseudo boundary. The Laplace's equation in the 2D domain Ω with a generalized boundary condition on the boundary Γ are expressed as follows:

$$\nabla^2 \phi = 0 \quad \text{in} \quad \Omega, \tag{1}$$

$$f_1 \phi + f_2 \frac{\partial \phi}{\partial n} = f_3 \quad \text{on} \quad \Gamma, \tag{2}$$

where, f_1, f_2 and f_3 are known functions on the boundary, n is the outward direction normal to the boundary Γ and ϕ is the primary variable of the problem. As previously mentioned, the solution is approximated by a linear combination of fundamental solutions as follows [14]:

$$\phi(\boldsymbol{x}) = \sum_{i=1}^{N} a_i \phi^*(\boldsymbol{x}, \boldsymbol{\xi}_i), \tag{3}$$

where a_i and $\boldsymbol{\xi}_i$ represent the intensity and location of the ith source located on the pseudo-boundary Γ', N is the number of source points, and $\boldsymbol{x}$ is the coordinate of the point in the domain or on the boundary of the solution domain. The constants a_i are the unknowns of the problem that have to be found. The fundamental solution for 2D Laplace equation can be expressed as follows [41]:

$$\phi^*(\boldsymbol{x}, \boldsymbol{\xi}_i) = -\frac{1}{2\pi} \ln r_i, \tag{4}$$

where r_i represents the distance between the source point $\boldsymbol{\xi}_i$ and the field point $\boldsymbol{x}$. Substituting eqn (4) into eqn (3) leads to:

$$\sum_{i=1}^{N} a_i \left[f_1(\boldsymbol{C_j}) \phi^*(\boldsymbol{C_j}, \boldsymbol{\xi}_i) + f_2(\boldsymbol{C_j}) \frac{\partial \phi^*(\boldsymbol{C_j}, \boldsymbol{\xi}_i)}{\partial n} \right] = f_3(\boldsymbol{C_j}), \quad j = 1, 2, ..., M, \tag{5}$$

where $\boldsymbol{C_1}, \boldsymbol{C_2}, ..., \boldsymbol{C_M}$ are collocation points. In order to find the unknowns of the problem i.e., a_i, the number of the equations or collocation points should be greater than the number of unknowns or source points ($M \geq N$). Eqn (5) indicates a system of M linear equations with N unknowns, which can be written in following form:

$$\mathbf{BX} = \mathbf{F}, \tag{6}$$

where the elements of the matrix $\mathbf{B} \in (\mathbf{R}^{M \times N}$, the vectors $\mathbf{X} \in \mathbf{R}^N$ and $\mathbf{F} \in \mathbf{R}^M$ can be expressed as:

$$B_{ji} = f_1(\mathbf{C_i})\phi^*(\mathbf{C_i}, \boldsymbol{\xi_j}) + f_2(\mathbf{C_j})\frac{\partial \phi^*(\mathbf{C_i}, \boldsymbol{\xi_j})}{\partial n}, \tag{7}$$

$$F_i = f_3(\mathbf{C_i}), \quad X_i = a_i. \tag{8}$$

In the case of $M = N$, the system of equations can be solved using standard methods such as the Gaussian elimination method or using the inverse of the coefficient matrix as follows:

$$\mathbf{X} = \mathbf{B}^{-1}\mathbf{F}. \tag{9}$$

In the case of $M > N$, the system of equations will be over-determined and can be solved in the least-squares sense as follows:

$$\mathbf{X} = (\mathbf{B}^T\mathbf{B})^{-1}\mathbf{B}^T\mathbf{F}. \tag{10}$$

When the intensity of the source points a_i are determined, the solution can be calculated on any internal or boundary points using eqn (3).

2.2 The direct BEM for Laplace's equation

The BEM is based on the Green theorem where the fundamental solution of the Laplace equation is taken as the auxiliary function. Two scalar functions ϕ^* and ϕ which are continuous over the domain boundary are considered in the BEM that correspond to the fundamental solution of the Laplace equation and the variable of the problem, respectively. Using the Green theorem, the integral equation for the Laplace's equation can be expressed as follows [41]:

$$\int_\Omega (\phi(\mathbf{z})\nabla^2\phi^*(\mathbf{x}, \mathbf{z}) - \phi^*(\mathbf{x}, \mathbf{z})\nabla^2\phi(\mathbf{z}))dv(\mathbf{z}) = \int_\Gamma (\phi(\mathbf{y})\frac{\partial \phi^*}{\partial n}(\mathbf{x}, \mathbf{y})$$
$$-\phi^*(\mathbf{x}, \mathbf{y})\frac{\partial \phi}{\partial n}(\mathbf{y}))ds(\mathbf{y}), \quad \mathbf{x}, \mathbf{z} \in \Omega, \mathbf{y} \in \Gamma. \tag{11}$$

The function ϕ corresponds to the variables of the problem defined on the domain Ω, ϕ^* is fundamental solution of the Laplace equation which is defined according to eqn (4) and n is the outward normal direction to the boundary. The point $\mathbf{x}$ in this equation corresponds to the location of a unit load in the auxiliary problem, while $\mathbf{y}$ and z are arbitrary points on or within the boundary of the problem. Eqn (11) can be simplified by using the Green theorem and integration techniques as follows [41]:

$$c(\mathbf{x})\phi(\mathbf{x}) + \int_\Gamma \phi(\mathbf{y})\frac{\partial \phi^*}{\partial n}(\mathbf{x}, \mathbf{y})ds(\mathbf{y}) = \int_\Gamma \phi^*(\mathbf{x}, \mathbf{y})\frac{\partial \phi}{\partial n}\mathbf{y})ds(\mathbf{y}), \quad \mathbf{x}, \mathbf{y} \in \Gamma. \tag{12}$$

$c(\mathbf{x})$ is a coefficient which presents the free term of integral equation and it will be calculated according to the geometry and boundary condition of the problem [41].

2.3 The MFS for 2D elastostatic problems

When there are no body forces, the governing equations for the elastostatic problem in 2D domain Ω and its boundary Γ are expressed as follows [42]:

$$\sigma_{ij,j} = 0, \tag{13}$$

$$\varepsilon_{ij} = \frac{1}{2}\left(u_{i,j} + u_{j,i}\right), \tag{14}$$

$$\sigma_{ij} = \frac{\nu E}{(1+\nu)(1-2\nu)}\delta_{ij}\varepsilon_{kk} + \frac{E}{1+\nu}\varepsilon_{ij}, \tag{15}$$

where σ_{ij} and ε_{ij} are the stress and strain tensors, respectively; u_i is the displacement vector, ν and E are the Poisson's ratio and Young module of the problem. The boundary condition for general 2D elastostatic problems can be expressed as:

$$f_{sj}u_j + g_{si}\sigma_{ij}n_j = p_s, \quad i,j,s = 1,2 \qquad \text{for 2D problems.} \tag{16}$$

f_{sj}, g_{si} and p_s are given functions on the boundary and n_j represent components of the outward unit vector normal to the boundary Γ. In the MFS for elastostatic problems, displacement components are generally explained as follows [14]:

$$u_i(\boldsymbol{x}) = \sum_{k=1}^{N}\sum_{m=1}^{2} a_m(\boldsymbol{\xi_k})u_{im}^*(\boldsymbol{x}, \boldsymbol{\xi_k}), \tag{17}$$

where ξ_k is the location of the kth source (concentrated fictitious force) located on the pseudo-boundary Γ', N is the number of source points, and $\boldsymbol{x}$ is a point in the domain or on the boundary of the solution domain. The constants $a_m(\xi_i)$ are the unknowns of the problem, which must be determined. The strain tensor components are obtained by substituting eqn (17) into (14), which yields:

$$\varepsilon_{ij}(\boldsymbol{x}) = \sum_{k=1}^{N}\sum_{m=1}^{2} a_m(\boldsymbol{\xi_k})\varepsilon_{ijm}^*, \tag{18}$$

where

$$\varepsilon_{ijm}^* = \frac{u_{im,j}^* + u_{jm,i}^*}{2}. \tag{19}$$

In addition, the stress tensor components may also be computed by substituting eqn (18) into eqn (15):

$$\sigma_{ij}(\boldsymbol{x}) = \sum_{k=1}^{N}\sum_{m=1}^{2} a_m(\boldsymbol{\xi_k})\sigma_{ijm}^*, \tag{20}$$

where

$$\sigma_{ijm}^* = \left[\frac{\nu E}{(1+\nu)(1-2\nu)}\delta_{ij}u_{tm,t}^* + \frac{E}{2(1+\nu)}(u_{im,j}^* + u_{jm,i}^*)\right]. \tag{21}$$

To calculate the stress and strain components, the first derivatives u_{ij}^* must also be calculated. The fundamental solution for 2D elasticity problems is given as follows [22]:

$$u_{ij}^* = \frac{1}{8\pi G(1-\bar{\nu})}\left[(3-4\bar{\nu})\delta_{ij}ln\frac{1}{r} + r_{,i}.r_{,j}\right], \quad i,j = 1,2. \tag{22}$$

The term r refers to the distance between source and any point within the domain or on its boundary and G is the shear module. $\bar{\nu}$ for plane strain problems is the same as ν while for plane stress problems is $\frac{\nu}{(1+\nu)}$. Differentiating eqn (22) leads to:

$$u_{ij,k}^* = \frac{1}{8\pi G(1-\bar{\nu})}\left[-(3-4\bar{\nu})\frac{r_k}{r^2}\delta_i j + \frac{r_j}{r^2}\delta_j k - \frac{2r_i r_j r_k}{r^4}\right], \quad i,j,k = 1,2, \quad (23)$$

where

$$r_i = x_i - \xi_i, r_{,i} = \frac{\partial r}{\partial x_i} = \frac{r_i}{r}, \qquad r = \sqrt{(x_1-\xi_1)^2 + (x_2-\xi_2)^2}, \qquad (24)$$

$$\xi_2 = \eta\xi_1 = \xi, \quad x_2 = y, \quad x_1 = x, \qquad (25)$$

$$r = \sqrt{(x-\xi)^2 + (y-\eta)^2}. \qquad (26)$$

After substituting eqns (17) and (20) into (16) for an arbitrary boundary point M, the following equation is obtained:

$$\sum_{k=1}^{N}\sum_{m=1}^{2} a_m(\boldsymbol{\xi_k})[f_{sj}(\boldsymbol{C_l})u_{jm}^*(\boldsymbol{C_l},\boldsymbol{\xi_k}) + g_{si}(\boldsymbol{C_l})\sigma_{ijm}^*(\boldsymbol{C_l},\boldsymbol{\xi_k})n_j] = p_s(\boldsymbol{C_l}),$$

$$l = 1,2,...,M. \qquad (27)$$

According to eqn (27), for 2D case, there will be 2M equations with 2N unknowns, in which M and N are collocation and source points, respectively. $C_1, C_2, ..., C_M$ in eqn (27) are collocation points. In order to obtain the unknowns $a_m(\xi_i)$, the number of equations should be equal or greater than the unknowns ($2N \geq 2M$). Eqn (27), which shows the system of 2M equations with 2N unknowns, can be written in the form of:

$$\mathbf{BX} = \mathbf{F}. \qquad (28)$$

When $M = N$, the system of equations in eqn (24) can be solved as $\mathbf{X} = \mathbf{B}^{-1}\mathbf{F}$. In the case of $M > N$ the system of equations should be solved using least-square method as $\mathbf{X} = (\mathbf{B}^{\mathbf{T}}\mathbf{B})\mathbf{B}^{\mathbf{T}}\mathbf{F}$. By calculating the coefficient of sources, solution in any arbitrary point inside the domain or on the boundary can be achieved using eqn (22).

2.4 The direct BEM for 2D elastostatic problems

The reciprocal theorem is used to derive integral equations of elasticity. Assume that an elastic body is subjected to two different loads and the solutions of the two problems are numbered (1) and (2), respectively. This theorem indicates that the work done by the first system of work acting via displacement of the second system is equal to the work done by the second system of forces acting via displacement of the first system, that is [42]:

$$\int_{\Gamma} t_i^{(1)}u_i^{(2)}d\Gamma + \int_{\Omega} F_i^{(1)}u_i^{(2)}d\Omega = \int_{\Gamma} t_i^{(2)}u_i^{(1)}d\Gamma + \int_{\Omega} F_i^{(2)}u_i^{(1)}d\Omega, \qquad (29)$$

where F_i and t_i represent body force and traction components, respectively. Using the reciprocity theorem, one can obtain the elasticity integral equation. This is done by considering the main problem as the first system and the Kelvin problem with the fundamental solution as the second one. The integral equation of elasticity in the absence of body force and in a boundary form can be written as follows [41]:

$$c_{ji}(\boldsymbol{x})u_i(\boldsymbol{x}) + \int_\Gamma u_{ij}^*(\boldsymbol{x}, \boldsymbol{y})u_i(\boldsymbol{y})ds(\boldsymbol{y}) = \int_\Gamma T_{ij}^*(\boldsymbol{x}, \boldsymbol{y})t_i(\boldsymbol{y})ds(\boldsymbol{y}), \;\; \boldsymbol{x}, \boldsymbol{y} \in \Gamma, \quad (30)$$

where u_{ij}^* represents the displacement fundamental solution of elasticity and is given in eqn (22) and c_{ji} are free term coefficients [43]. T_{ij}^* is expressed as follows [41]:

$$T_{ij}^* = \frac{-1}{4\pi(1-\bar{\nu})r}\left[\frac{\partial r}{\partial n}[(1-2\bar{\nu})\delta_{ij} + 2r_{,i}.2r_{,j}] - (1-2\bar{\nu})(r_in_j - r_jn_i)\right]. \quad (31)$$

The component of stress after discretizing the boundary of the domain with a number of boundary elements and after some numerical manipulations, the discretized form of the boundary element equation may be written in matrix form as:

$$Hu = Gt. \quad (32)$$

After satisfying the prescribed boundary conditions and rearrangement of unknowns (displacement or traction) a standard system of algebraic equations is obtained that can be solved by standard system solving methods.

2.5 A suitable configuration for source points in the MFS for 2D elasticity

The arrangement of source and collocation points has considerable effects on the accuracy of the MFS results. In some problems, a little change in the configuration of source points may lead to a considerable error in the results. Although extensive studies have been conducted on the location of source points in the MFS, investigation on this topic is still underway. In this work, we select the location of source points based on the recommendations provided by Hematiyan et al. [3], [31]. According to the suggestion of Hematiyan et al. [3], the positions of source points in the 2D potential problem is determined in such a way that the value of the location parameter of the source points would be greater than 0.8. The value of the location parameter in the 2D elastostatic problem is recommended to be greater than 0.85. The value of the location parameter determines the location of an individual source point relative to its neighboring source points and the boundary of the problem. The location parameter can be defined using Fig. 1, which shows the schematic view of the main boundary, the pseudo boundary and source points. The closest boundary point to the ith source point $\mathbf{S}_i$ is considered as its base point and is donated by $\mathbf{B}_i$.

The location parameter is defined as follows:

$$K_i = \frac{d_{i/i}}{max(d_{i-1/i}, d_{i+1/i})}, \quad 0 < K_i < 1. \quad (33)$$

where $d_{i/j}$ represents the distance between the base point $\mathbf{B}_j$ to the source point $\mathbf{S}_j$. If the location parameter has a small value (the source points are close to the boundary), the solutions will have an oscillation, while a larger value (near to 1.0) of the location parameter (source points are far from the boundaries) will result in ill-conditioned system of equations.

3 RESULTS AND DISCUSSION

In this section by presenting several numerical examples with different geometries and boundary conditions, the accuracy of the BEM and the MFS for convex and concave domains are compared.

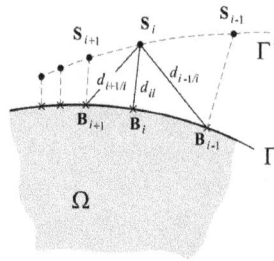

Figure 1: The location of a source point relative to the base points of the neighboring source points [3].

Table 1: Comparing temperature values in BEM and MFS for a Dirichlet boundary condition for a rectangular sheet.

Node numbers	T_{MFS}	T_{BEM}
24	2.17662	2.286
54	2.1755	2.179
100	2.1755	2.176
Reference solution	2.1755	2.1755

3.1 Comparison of the BEM and the MFS for potential problems

3.1.1 A rectangular plate with Newmann boundary condition

A rectangular domain (2×1) is considered with a coordinate system located at the center of the rectangle. There are the following boundary conditions for the Laplace equation to be solved:

$$\phi = x^2 + y^2 \qquad \text{on} \quad \Gamma. \tag{34}$$

The problem has been solved using BEM and MFS with 24, 56, and 100 nodes. The MFS configuration for the three cases is shown in Fig. 2. All three of these cases show a magnitude of 0.9 for the location parameter. In the boundary element nodes are located on the boundary at the same distance. A reference point of (1.25,0) is specified between BEM and MFS for comparison purposes. A reference solution was obtained using the finite element method with appropriate grid numbers. Table 1 provides the results for all three methods with different node numbers. According to a reference solution, both methods are capable of providing a reliable solution, and increasing the number of nodes will increase accuracy. While both methods are acceptable in accuracy, it's apparent that the MSF is more accurate with a lower number of nodes.

3.1.2 Sheet plate with concave domain

This example illustrates how geometrical complexity of the domain affects the accuracy of the solution obtained from MFS and BEM. Fig. 3 illustrates the plate with a concave domain and boundary conditions. The boundary conditions and the governing equation of the sheet

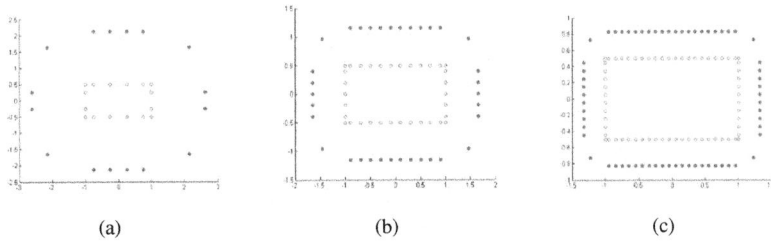

Figure 2: MFS arrangement of source and collocation points for a rectangular sheet with the number of (a) 24; (b) 54; (c) 100 nodes.

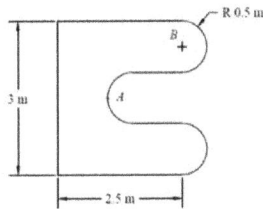

Figure 3: Concave sheet.

Table 2: A comparison of temperature and flux values at Point A in a concave sheet domain.

Node numbers	T_{MFS}	T_{BEM}	$TG_{x(BEM)}$	$TG_{x(MFS)}$
35	3.250	3.249	1.0458	1.0245
67	3.250	3.249	1.0267	1.0270
90	3.250	3.250	1.0250	1.0285
Reference solution	3.250	3.250	1.0275	1.0275

are as follows:

$$T = x^2 + y^2 \qquad \text{on the boundary,} \tag{35}$$

$$\nabla^2 T = 0 \qquad \text{inside the domain.} \tag{36}$$

Temperature and flux at two arbitrary points A (1, 1.5) and B (2.5, 2.5) are calculated in order to compare the accuracy of the proposed methods. The accuracy of the problem is examined using 35, 67, and 90 nodes. Based on these values, BEM and MFS solutions were compared. Finite element software is used to obtain reference values for flux and temperature at points A and B. Tables 2 and 3 provide the calculated values of temperature and flux at points A and B.

For BEM (Fig. 4), all nodes are located at the same distance from the boundary, while for MFS (Fig. 5), the distance between nodes at the corner is smaller than for other nodes, this distance will be achieved using location parameter.

WIT Transactions on Engineering Sciences, Vol 135, © 2023 WIT Press
www.witpress.com, ISSN 1743-3533 (on-line)

Table 3: A comparison of temperature and flux values at Point B in a concave sheet domain.

Node numbers	T_{MFS}	T_{BEM}
35	12.875	12.865
67	12.871	12.868
90	12.870	12.868
Reference solution	12.870	12.870

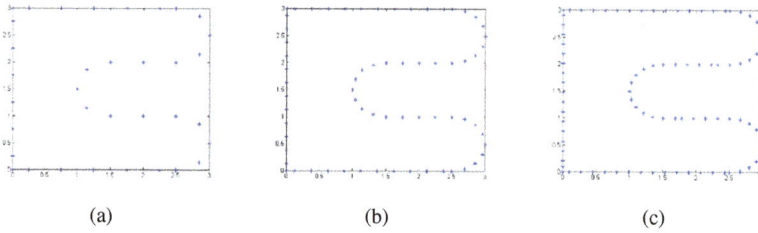

(a) (b) (c)

Figure 4: A BEM with (a) 35; (b) 67; and (c) 90 nodes for concave sheet plate.

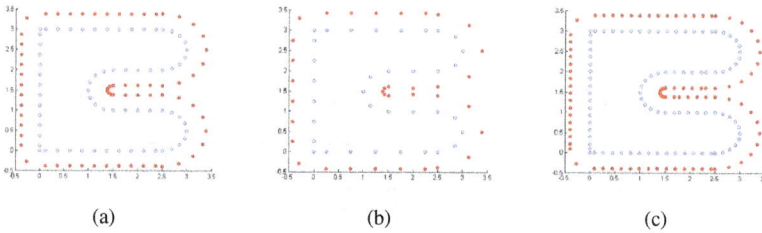

(a) (b) (c)

Figure 5: MFS arrangement of source points and collocation points with (a) 35; (b) 67; and (c) 90 nodes.

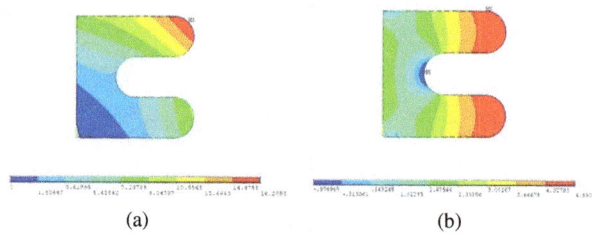

(a) (b)

Figure 6: (a) Temperature contours; and (b) Flux contours in concave sheet plate in FEM.

A reference solution for temperature and flux is calculated and simulated using finite element software. Figs 6 and 7 represent the temperature and flux contours, respectively.

It has been necessary to calculate the temperature and flux contours obtained from MFS in order to gain a more comprehensive understanding of the results. The contours of temperature

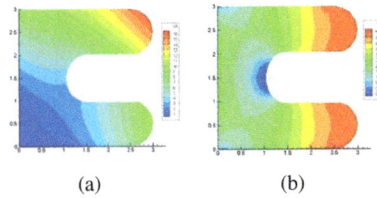

(a)   (b)

Figure 7: (a) Temperature contours; and (b) Flux contours in concave sheet plate in MFS.

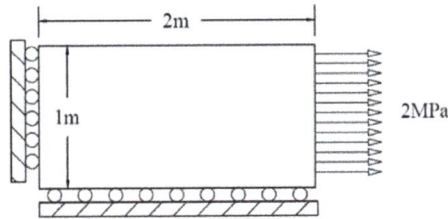

Figure 8: The rectangular plate with plane strain condition.

and flux are shown in Fig. 8. As can be seen, the solution calculated by MFS on the domain of the problem is similar to the reference solution calculated by finite element software.

Clearly, both BEM and MFS can provide high accuracy results for both temperature and flux fields. The accuracy of both methods is the same, and the convergence rate to the reference solution is increased by increasing the number of nodes. The FEM method requires the use of thousands of nodes for fine-meshing, while the MFS and BEM methods can achieve high accuracy results with only 90 nodes. As an alternative to MFS, BEM can provide more accurate solutions in a concave domain, not only at point A, but also at internal point B. As a result, it can be concluded that the arrangement of source points and collocation points in MFS is critical. It is preferred to use BEM in cases of geometric complexity in which a suitable arrangement of source points and collocation points cannot be achieved.

3.2 Comparison of BEM and MFS for elasticity problems

3.2.1 Rectangular plate with plane strain condition

In this study, a rectangular sheet plate with roller support on two sides was considered, and the results of BEM and MFS were compared when a distributed tension load of 2 MPa was applied. Fig. 8 illustrates the geometry, loading, and boundary conditions of the problem. In solving the problem, Young's module ($E = 2 \times 10^9$) and Poisson's ratio ($\nu = 0.3$) were considered. Different numbers of source points were used to solve the problem, namely 24, 48, and 96. As a result of the study, error percentages for each method were calculated. It is necessary to determine the appropriate location for source points and collocation points. Based on evaluations, the location parameter of 0.85 has been chosen. Fig. 9 illustrates the configuration for source points and collocation points in MFS. In BEM, the arrangements of boundary nodes are presented in Fig. 10 with different numbers of boundary nodes. We should remember that despite MFS, nodes can be positioned at the corner of a rectangular

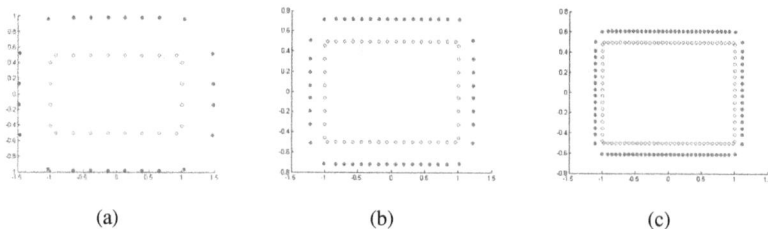

Figure 9: Arrangement of source points and collocation points for a rectangular plate with plane strain condition in MFS with number of (a) 24; (b) 48; and (c) 98 nodes.

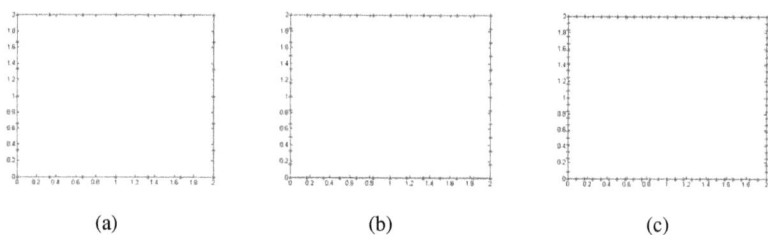

Figure 10: Arrangement boundary node for a rectangular plate with plane strain condition in BEM with number of (a) 24; (b) 48; and (c) 98 nodes.

Table 4: Percentage of displacement error in MFS and BEM for rectangular sheet.

Node numbers	$E_{ux(MFS)}$	$E_{ux(BEM)}$	$E_{uy(MFS)}$	$E_{uy(BEM)}$
24	23.6×10^{-4}	0.26×10^{-4}	17.17×10^{-4}	0.24×10^{-4}
48	20.8×10^{-4}	0.18×10^{-4}	10.9×10^{-4}	0.146×10^{-4}
96	1.28×10^{-4}	0.13×10^{-4}	0.43×10^{-4}	0.0096×10^{-4}

plate in BEM as well. In the case of plain stress conditions, the displacement formulation provides an accurate solution.

As shown in Table 4, the mean error is calculated for each method with different numbers of nodes, the indices ux and uy indicate the horizontal and vertical displacements, respectively. The results indicate that both methods provide accurate and precise answers. With an increase in the number of nodes, the accuracy of the solutions will also increase. Despite the fact that the rate of error for the two methods is practically negligible, the accuracy or robustness of the BEM method has a distinct advantage over the fundamental solution method. solution method.

3.2.2 A plate with a deep hole
Following the previous problem, the hole within the sheet has deepened, and the stress values at points A and B will be analyzed. Fig. 11 presents the boundary conditions and geometry of the example.

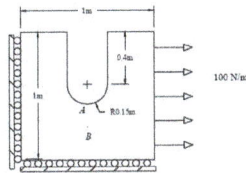

Figure 11: Plate with much deeper hole.

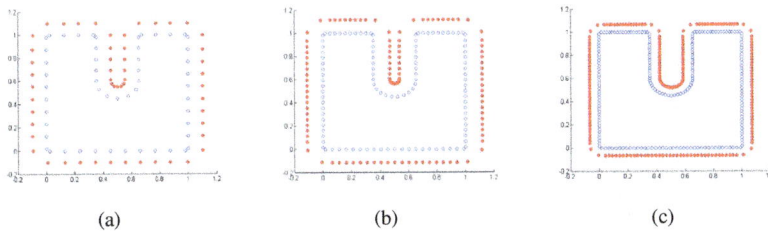

| (a) | (b) | (c) |

Figure 12: Arrangement of source points and collocation points for a plate with a deep hole with number of (a) 50; (b) 111; and (c) 213 nodes in MFS.

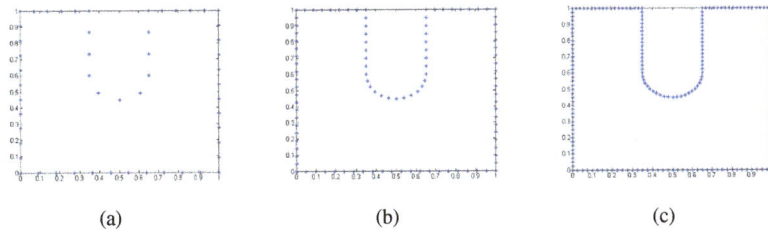

| (a) | (b) | (c) |

Figure 13: Arrangement of boundary nodes for a plate with a deep hole with number of (a) 50; (b) 111; and (c) 213 nodes in BEM.

Based on finite element software, a reference solution of stresses at points A and B has been obtained. According to MFS, the appropriate arrangement of source and collocation points yields a position parameter of 0.85 approximately. To solve this example, both methods consider three distinct node arrangements, namely 53, 111, and 213 nodes. Figs 12 and 13 illustrate the arrangement of points in the BEM and MFS respectively. In MFS, it should be noted that no nodes are placed in corners. Tables 5 and 6 provide a comparison between BEM and MFS results at point A and B. In order to determine the efficacy of BEM, two types of elements were used namely linear and quadratic.

It can be seen from Tables 5 and 6 that BEM has superiority over MFS, particularly when considering quadratic elements. Several configurations were examined to determine the effect of the location parameter on the results accuracy. The location parameter plays a significant role in determining the stress at point A, as shown in Table 7, in MFS a slight change in this parameter will have a significant impact on the results. Based on the results of this example, it can be concluded that MFS is less efficient than BEM, especially with a small number

Table 5: Comparison of stress values for a plate with much deeper hole at point A.

Node	$\sigma_{x(BEM)}$ Linear	Quadratic	$\sigma_{x(MFS)}$
53	570.0	901.2	496.27
111	762.7	765.2	749.0
213	780.2	780.5	786.3
Reference solution	768.3	768.3	786.3

Table 6: Comparison of stress values for a plate with a deep hole at point B.

Node	$\sigma_{x(BEM)}$ Linear	Quadratic	$\sigma_{x(MFS)}$	$\sigma_{y(BEM)}$ Linear	Quadratic	$\sigma_{y(MFS)}$	$\tau_{xy(BEM)}$ Linear	Quadratic	$\tau_{xy(MFS)}$
53	131.2	135.06	120.33	138.7	152.10	62.98	59.9	67.8	29.96
111	135.6	135.96	134.50	149.1	150.52	140.5	67.3	68.4	64.00
213	135.8	135.98	135.98	150.1	150.50	148.9	68.1	68.3	68.10
Reference	139.6	139.6	139.6	152.5	152.5	152.5	71	71	71

Table 7: Comparison of stress values of a plate with much deeper hole at point A for different location parameter values.

Location parameter of source points	$\sigma_{x(MFS)}$
0.89	63.93
0.85	538.69
0.82	107.98
0.80	438.23
0.70	638.35
0.67	598.12
0.64	280.41

of collocation and source points where it may be physically impossible to achieve a proper arrangement satisfying the location parameters constraint. However, BEM does not have this limitation and is therefore able to provide more accurate results in this regard, particularly when quadratic elements are used.

4 CONCLUSION

In this study, BEM and MFS were compared with regard to potential problems and elasticity problems. The efficiency of the two proposed methods was assessed over a variety of boundary domains, including simple and concave boundaries. The major challenge facing MFS is to arrange source points and collocation points in such a way that reliable results will be obtained. A location parameter was defined to determine the optimal location of source points relative to each other and the boundary of the problem. According to the results, both methods provide accurate results for simple geometry. In complex geometries, such as in concave domains (where stress concentration is high), the MFS is not as effective as it should be. With such a small number of source points, it is not possible to satisfy the location parameter and reach the desired location parameter. Therefore, the results of the analysis are unreliable. Despite this, BEM has the potential to provide accurate results, even if the geometry of the problem is complex, and quadratic elements will enhance this accuracy. Since input data preparation in both proposed methods occur only at the boundary of the problem, it can be concluded that BEM and MFS reduce simulation time and input data compared to domain types of the problem such as the finite element method.

REFERENCES

[1] Yokoyama, M. & Kanoh, T., Accuracy estimation of the three-D BEM analysis of elastostatic problems by the zooming method. *Advances in Engineering Software*, **31**, pp. 355–365, 2000.

[2] Cheng, A.H. & Hong, Y., An overview of the method of fundamental solutions: Solvability, uniqueness, convergence, and stability. *Engineering Analysis with Boundary Elements*, pp. 118–152, 2020.

[3] Hematiyan, M., Haghighi, A. & Khosravifard, A., A two-constraint method for appropriate determination of the configuration of source and collocation points in the method of fundamental solutions for 2D Laplace equation. *Advances in Applied Mathematics and Mechanics*, 2017.

[4] Drombosky, T.W., Meyer, A.L. & Ling, L., Applicability of the method of fundamental solutions. *Engineering Analysis with Boundary Elements*, **33**(5), pp. 637–643, 2009.

[5] Askour, O., Tri, A., Braikat, B., Zahrouni, H. & Potier-Ferry, M., Method of fundamental solutions and high order algorithm to solve nonlinear elastic problems. *Engineering Analysis with Boundary Elements*, **89**, pp. 25–35, 2018.

[6] Wang, F., Fan, C.M., Hua, Q. & Gu, Y., Localized MFS for the inverse Cauchy problems of two-dimensional Laplace and biharmonic equations. *Applied Mathematics and Computation*, **364**, 124658, 2020.

[7] Jaswon, M., Integral equation methods in potential theory. I. *Proceedings of the Royal Society of London Series A Mathematical and Physical Sciences*, **275**(1360), pp. 23–32, 1963.

[8] Symm, G., Integral equation methods in potential theory. II. *Proceedings of the Royal Society of London Series A Mathematical and Physical Sciences*, **275**(1360), pp. 33–46, 1963.

[9] Rizzo, F. & Shippy, D., A boundary integral approach to potential and elasticity problems for axisymmetric bodies with arbitrary boundary conditions. *Mechanics Research Communications*, **6**(2), pp. 99–103, 1979.

[10] Muskhelishvili, N. et al., *Some Basic Problems of the Mathematical Theory of Elasticity*, vol. 15. Noordhoff Groningen, 1953.

[11] Kupradze, V., Potential methods in elasticity theory. Technical report, Foreign Technology Div Wright-Patterson AFB Ohio, 1967.

[12] Kupradze, V.D. & Aleksidze, M.A., The method of functional equations for the approximate solution of certain boundary value problems. *USSR Computational Mathematics and Mathematical Physics*, **4**(4), pp. 82–126, 1964.

[13] Mathon, R. & Johnston, R.L., The approximate solution of elliptic boundary-value problems by fundamental solutions. *SIAM Journal on Numerical Analysis*, **14**(4), pp. 638–650, 1977.

[14] Karageorghis, A. & Fairweather, G., The method of fundamental solutions for the solution of nonlinear plane potential problems. *IMA Journal of Numerical Analysis*, **9**(2), pp. 231–242, 1989.

[15] Fam, G.S. & Rashed, Y.F., The method of fundamental solutions applied to 3D structures with body forces using particular solutions. *Computational Mechanics*, **36**(4), pp. 245–254, 2005.

[16] Young, D., Tsai, C.C., Chen, C. & Fan, C.M., The method of fundamental solutions and condition number analysis for inverse problems of Laplace equation. *Computers and Mathematics with Applications*, **55**(6), pp. 1189–1200, 2008.

[17] Marin, L., Karageorghis, A. & Lesnic, D., The MFS for numerical boundary identification in two-dimensional harmonic problems. *Engineering Analysis with Boundary Elements*, **35**(3), pp. 342–354, 2011.

[18] Karageorghis, A., Lesnic, D. & Marin, L., A survey of applications of the MFS to inverse problems. *Inverse Problems in Science and Engineering*, **19**(3), pp. 309–336, 2011.

[19] Fan, C.M. & Li, P.W., Numerical solutions of direct and inverse stokes problems by the method of fundamental solutions and the Laplacian decomposition. *Numerical Heat Transfer, Part B: Fundamentals*, **68**(3), pp. 204–223, 2015.

[20] Mohammadi, M., Hematiyan, M.R. & Shiah, Y., An efficient analysis of steady-state heat conduction involving curved line/surface heat sources in two/three-dimensional isotropic media. *Journal of Theoretical and Applied Mechanics*, **56**(4), pp. 1123–1137, 2018.

[21] Gorzelańczyk, P. & Kołodziej, J.A., Some remarks concerning the shape of the source contour with application of the method of fundamental solutions to elastic torsion of prismatic rods. *Engineering Analysis with Boundary Elements*, **32**(1), pp. 64–75, 2008.

[22] Berger, J.R. & Karageorghis, A., The method of fundamental solutions for layered elastic materials. *Engineering Analysis with Boundary Elements*, **25**(10), pp. 877–886, 2001.

[23] Karageorghis, A., A practical algorithm for determining the optimal pseudo-boundary in the method of fundamental solutions. *Adv. Appl. Math. Mech.*, **1**(4), pp. 510–528, 2009.

[24] Li, M., Chen, C. & Karageorghis, A., The MFS for the solution of harmonic boundary value problems with non-harmonic boundary conditions. *Computers and Mathematics with Applications*, **66**(11), pp. 2400–2424, 2013.

[25] Alves, C.J., On the choice of source points in the method of fundamental solutions. *Engineering Analysis with Boundary Elements*, **33**(12), pp. 1348–1361, 2009.

[26] Liu, C.S., An equilibrated method of fundamental solutions to choose the best source points for the Laplace equation. *Engineering Analysis with Boundary Elements*, **36**(8), pp. 1235–1245, 2012.

[27] Wang, F., Liu, C.S. & Qu, W., Optimal sources in the MFS by minimizing a new merit function: Energy gap functional. *Applied Mathematics Letters*, **86**, pp. 229–235, 2018.

[28] Cisilino, A.P. & Sensale, B., Application of a simulated annealing algorithm in the optimal placement of the source points in the method of the fundamental solutions. *Computational Mechanics*, **28**(2), pp. 129–136, 2002.

[29] Jopek, H. & Kołodziej, J., Application of genetic algorithms for optimal positions of source points in the method of fundamental solutions. *Comput. Assist. Mech. Eng. Sci.*, **15**(3–4), pp. 215–224, 2008.

[30] Chen, C., Karageorghis, A. & Li, Y., On choosing the location of the sources in the MFS. *Numerical Algorithms*, **72**, pp. 107–130, 2016.

[31] Hematiyan, M., Arezou, M., Dezfouli, N.K. & Khoshroo, M., Some remarks on the method of fundamental solutions for two-dimensional elasticity. *Comput. Model Eng. Sci. (CMES)*, **121**, pp. 661–686, 2019.

[32] Tadeu, A., Godinho, L. & Chen, C., Performance of the BEM, MFS, and RBF collocation method in a 2.5 d wave propagation analysis. *WIT Transactions on Modelling and Simulation*, vol. 39, WIT Press: Southampton and Boston, 2005.

[33] Salgado-Ibarra, E.A., Boundary element method (BEM) and method of fundamental solutions (MFS) for the boundary value problems of the 2-D Laplace's equation, 2011.

[34] Alves, C.J., Martins, N.F. & Valtchev, S.S., Trefftz methods with cracklets and their relation to BEM and MFS. *Engineering Analysis with Boundary Elements*, **95**, pp. 93–104, 2018.

[35] Ahmed, M.T., Lavers, J. & Burke, P., An evaluation of the direct boundary element method and the method of fundamental solutions. *IEEE Transactions on Magnetics*, **25**(4), pp. 3001–3006, 1989.

[36] Nath, D. & Kalra, M., Solution of Grad–Shafranov equation by the method of fundamental solutions. *Journal of Plasma Physics*, **80**(3), pp. 477–494, 2014.

[37] Godinho, L., Costa, E., Pereira, A. & Santiago, J., Some observations on the behavior of the method of fundamental solutions in 3D acoustic problems. *International Journal of Computational Methods*, **9**(04), 1250049, 2012.

[38] Godinho, L. & Soares Jr, D., Frequency domain analysis of interacting acoustic–elastodynamic models taking into account optimized iterative coupling of different numerical methods. *Engineering Analysis with Boundary Elements*, **37**(7–8), pp. 1074–1088, 2013.

[39] Dyhoum, T., Lesnic, D. & Aykroyd, R., Solving the complete-electrode direct model of ERT using the boundary element method and the method of fundamental solutions. *Electronic Journal of Boundary Elements*, **12**(3), 2014.

[40] Liravi, H., Arcos, R., Ghangale, D., Noori, B. & Romeu, J., A 2.5 D coupled FEM-BEM-MFS methodology for longitudinally invariant soil-structure interaction problems. *Computers and Geotechnics*, **132**, 104009, 2021.

[41] Paris, F. & Canas, J., *Boundary Element Method: Fundamentals and Applications*, Oxford University Press: USA, 1997.

[42] Sadd, M.H., *Elasticity: Theory, Applications, and Numerics*, Academic Press, 2009.

[43] Gao, X.W. & Davies, T.G., *Boundary Element Programming in Mechanics*, Cambridge University Press, 2002.

PROGRESS OF MESH REDUCTION METHODS

XIAO-WEI GAO, HUA-YU LIU & WEI-LONG FAN
State Key Laboratory of Structural Analysis for Industrial Equipment, Dalian University of Technology, China

ABSTRACT
According to the discretization manner and manipulation dimensionality, numerical methods can be classified into two big groups, the full dimensionality method and mesh reduction methods. The former includes the finite element method, finite block method, element differential method and so on, and the latter can be further divided into three types: surface-based, line-based and point-based methods. The surface-based method consists of the boundary element method, finite volume method, boundary face method and so on; the line-based method includes the finite difference method, finite element line method, finite line method and so on; and the point-based method mainly consists of the mesh free method, fundamental solution method, free element method and so on. The paper gives a detailed description on the progress achieved in recent years in the mesh reduction methods, among which a brief introduction will go into the surface- and point-based methods and a detailed one will be given to the line-based methods, especially the finite line method (FLM). FLM is a new type collocation method, which has a distinct feature that the solution scheme to solve a partial differential equation is established by using two or three lines crossing a collocation point for 2D or 3D problems. The Lagrange polynomial formulation is used to approximate the variation of the coordinates and physical variables over each line. And the directional derivative technique is used to derive various orders spatial partial derivatives of any physical variables. The derived partial derivatives can be directly substituted into the governing partial differential equations and related boundary conditions to set up the discretized system of equations. An example will be given for the mechanical problem to demonstrate the accuracy and efficiency of the descripted mesh reduction methods.
Keywords: mesh reduction method, finite line method, collocation method, finite difference method, mesh free method.

1 INTRODUCTION
Most engineering problems can be represented by a set of second-order partial differential equations (PDEs) with proper boundary conditions (BC), which are constituted of the so called boundary value problems (BVP) of PDEs [1]. For example, the solid mechanics problem has the following BVP [2]:

$$\text{PDE:} \qquad \frac{\partial}{\partial x_l}\left(D_{ijkl}(\boldsymbol{x})\frac{\partial u_k(\boldsymbol{x})}{\partial x_j} \right) + b_i(\boldsymbol{x}) = 0, \quad \boldsymbol{x} \in \Omega, \tag{1}$$

$$\text{BC:} \qquad \begin{cases} u_i(\boldsymbol{x}) = \bar{u}_i, & \boldsymbol{x} \in \Gamma_u \\ D_{ijkl}(\boldsymbol{x})n_j(\boldsymbol{x})\dfrac{\partial u_k(\boldsymbol{x})}{\partial x_l} = \bar{t}_i(\boldsymbol{x}), & \boldsymbol{x} \in \Gamma_t, \end{cases} \tag{2}$$

in which u_k is the displacement component, D_{ijkl} the stress–strain constitutive tensor, b_i the body force, and $\bar{t}_i$ the specified traction.

To solve the PDEs with a set of properly posed BC as listed above, a number of numerical methods are available, which can be globally divided into two big categories according to the geometry discretization and operation dimensions: the full dimensionality methods including the finite element method [3], finite block method [4], element differential method [2], and

WIT Transactions on Engineering Sciences, Vol 135, © 2023 WIT Press
www.witpress.com, ISSN 1743-3533 (on-line)
doi:10.2495/BE460091

so on, and mesh reduction methods including surface-based methods (i.e., the boundary element method [5], finite volume method [6], etc.), the line-based methods (i.e., the finite difference method [7], finite line method [8], etc.), and the point-based methods (i.e., the mesh free method [9], free element method [10], fundamental solution method [11], etc.). Most of these methods include two types of algorithms: weak-form and strong-form algorithms [9]. The weak-form algorithm needs integration over elements or divided sub-domains, and the strong-form algorithm usually is a type of collocation method without needing integration computation. In this paper, progresses in the mesh reduction methods are resumptively described.

A number of derivation techniques can be used to establish the solution schemes for above mentioned numerical methods. In view of the universal property, the weighted residual technique is used in the paper to establish the solution schemes for mesh reduction methods.

2 WEIGHTED RESIDUAL FORMULA FOR SOLVING PDES

Multiplying the PDE (1) on both sides with a weight function w and integrating it through the computational domain, it follows that:

$$\int_{\Omega} w(\boldsymbol{x})\frac{\partial}{\partial x_l}\left(D_{ijkl}(\boldsymbol{x})\frac{\partial u_k(\boldsymbol{x})}{\partial x_j}\right)d\Omega + \int_{\Omega} w(\boldsymbol{x})b_i(\boldsymbol{x})d\Omega = 0. \qquad (3)$$

Taking integration by parts and applying the Gauss' divergence theorem to the first domain integral of eqn (3), the above equation becomes:

$$\int_{\Omega} \frac{\partial w(\boldsymbol{x})}{\partial x_l}D_{ijkl}(\boldsymbol{x})\frac{\partial u_k(\boldsymbol{x})}{\partial x_j}d\Omega = \int_{\Gamma} w(\boldsymbol{x})t_i(\boldsymbol{x})d\Gamma + \int_{\Omega} w(\boldsymbol{x})b_i(\boldsymbol{x})d\Omega, \qquad (4)$$

in which t_i is the traction component on the boundary Γ of the domain Ω, which has the relationship with the displacement gradient as shown in eqn (2).

In eqn (4), the basic physical variable u_k is mainly included in the volume integral of the left-hand side, so it is called the volume-based weighted residual formulation.

Taking integration by parts again to the first domain integral of eqn (4) and applying the Gauss' divergence theorem, the following equation can be resulted in:

$$\int_{\Gamma} \frac{\partial w}{\partial x_l}D_{ijkl}n_j u_k d\Gamma - \int_{\Omega} \frac{\partial}{\partial x_j}\left(D_{ijkl}\frac{\partial w}{\partial x_l}\right)u_k d\Omega = \int_{\Gamma} wt_i d\Gamma + \int_{\Omega} wb_i d\Omega. \qquad (5)$$

In eqn (5), the basic physical variable u_k is included in both the surface and volume integrals of the left-hand side, so it is called the surface-volume-based weighted residual formulation.

From eqns (4) and (5), various weak-form solution algorithms can be generated by taking different kinds of the weight function w, such as the finite volume method and boundary element method.

3 SURFACE-BASED METHODS

The surface-based methods include the finite volume method (FVM) and boundary element method (BEM), which are operated mainly on the surfaces of a control volume or on the boundary of the considered problem.

3.1 Progress in FVM

The FVM seems to be a volume-based method. However, it is here classified into the category of the surface-based methods, because its main operation is over the surfaces rather than the volume. To see this, let us take the weight function w to be 1 in eqn (4) and thus it results in:

$$\int_{\Gamma} t_i \, d\Gamma + \int_{\Omega} b_i d\Omega = 0. \tag{6}$$

In FVM, the computational domain is discretized into a serious of control volumes [12]. Applying eqn (6) to each control volume, say volume Ω_c, and dividing the boundary of Ω_c into two parts, the inner and out boundary parts, eqn (6) can be written as:

$$\int_{\Gamma_{lc}} D_{ijkl} n_j \frac{\partial u_k}{\partial x_l} d\Gamma + \int_{\Gamma_{oc}} \bar{t_i} \, d\Gamma + \int_{\Omega_c} b_i d\Omega = 0, \tag{7}$$

where $\Gamma_{lc} \cup \Gamma_{oc} = \partial\Omega_c$.

Eqn (7) is a typical formulation in FVM analysis, from which we can see that the main computation is over the control surfaces of a control volume. The key work in FVM is how to compute the physical variable gradient $\partial u_k / \partial x_l$ [13] included in the first control surface integral of eqn (7).

3.1.1 Conventional FVM

In the conventional FVM, the interface Γ_{lc} is taken as the mid-surface connected by the collocation point c and around neighbour points, thus $\partial u_k / \partial x_l$ at the mid-surface can be easily computed by the values of u_k between c and the neighbour points [12], [13]. The point representing the physical variables may be the centre of a control volume (this being called the cell-centred FVM) or a vertex of an unstructured mesh (this being called the vertex-centred FVM) [14]. Figs 1 and 2 show two patterns of the vertex-centred FVM for a structured mesh and an unstructured mesh, respectively.

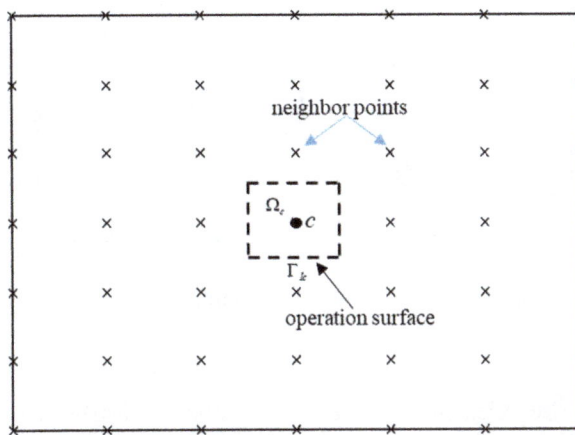

Figure 1:   Operation surface formed by mid-points between collocation point c and neighbour points for a structural mesh.

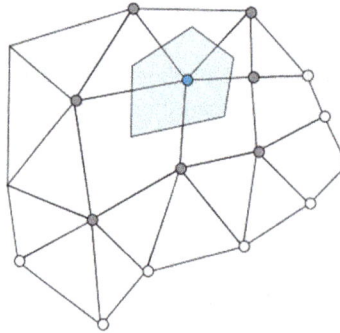

Figure 2: Control volume and its operation surface connected by mid-points between collocation point and neighbour vertices of an unstructured mesh [14].

Obviously, the linear variation of over the operation surface shown in Fig. 2 can be easily constructed, but it is difficult to construct a high-order scheme to compute the value of on the operation surface shown in Figs 1 and 2.

3.1.2 Free element based FVM (FEFVM)

In recent years, the authors proposed the free element method (FrEM) [10]. In the method, an isoparametric free element is defined at each collocation point as shown in Fig. 3. The weak-form formulation of FrEM has the form as shown in eqn (7), but a high-order control volume, the free element, can be easily formed as in FEM. The control surface in FEFVM is the boundary of the free element.

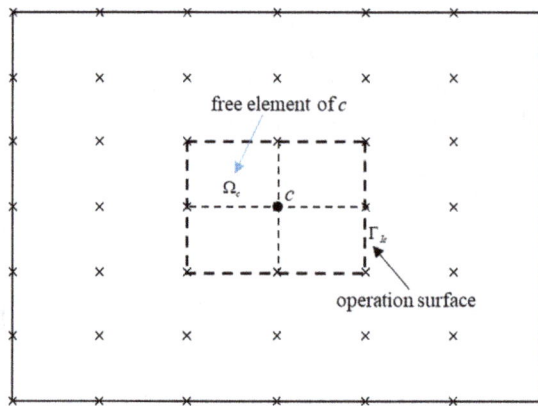

Figure 3: Operation surface formed by the boundary of the free element built for collocation point c.

Since high-order free elements can be easily formed in FrEM, the high accuracy of $\partial u_k / \partial x_l$ in FEFVM can also be easily achieved. A set of analytical expressions for computing $\partial u_k / \partial x_l$ over the formed free element was derived in Gao et al. [2]. Although it is easy to set up the solution scheme using the free element as shown in Fig. 3, the accuracy of $\partial u_k / \partial x_l$ is not high. This is because its value is taken on the boundary of Ω_c, which is not as accurate as inside Ω_c. To improve the accuracy of $\partial u_k / \partial x_l$, the element-shell strengthened FVM is proposed in the following.

3.1.3 Element-shell strengthened FVM (ESFVM)

To improve the accuracy of $\partial u_k / \partial x_l$ included in the first boundary integral of eqn (7), additional free elements are formed for each side/surface of the collocation element Ω_c, which form an element ring/shell for a 2D/3D control sides/surfaces as shown in Fig. 4 for a 2D case.

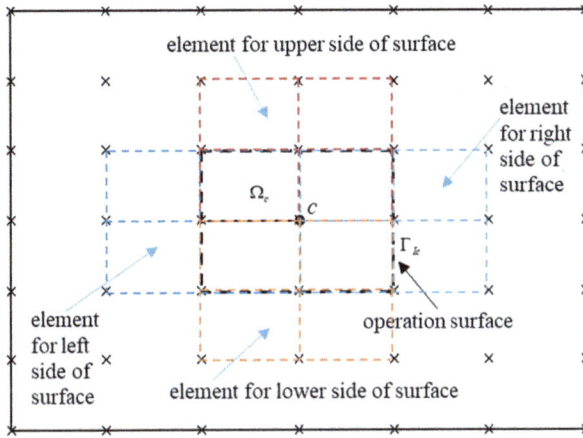

Figure 4: Additional elements around the operation surfaces.

In ESFVM, since the operation surfaces of Ω_c are included in the formed additional elements, the accuracy of $\partial u_k / \partial x_l$ is higher than using a same element as shown in Fig. (4).

3.2 Progress in BEM

In eqn (5), we take the weight function w as the displacement fundamental solution u_{mi}^* [5], that is $w = u_{mi}^*$, it follows that:

$$\int_{\Gamma} \frac{\partial u_{mi}^*}{\partial x_l} D_{ijkl} n_j u_k d\Gamma - \int_{\Omega} \frac{\partial}{\partial x_j} \left(D_{ijkl} \frac{\partial u_{mi}^*}{\partial x_l} \right) u_k d\Omega = \int_{\Gamma} u_{mi}^* t_i d\Gamma + \int_{\Omega} u_{mi}^* b_i d\Omega. \tag{8}$$

Because the fundamental solution u_{mi}^* satisfies the following equation:

$$\frac{\partial}{\partial x_j} \left(D_{ijkl} \frac{\partial u_{mi}^*}{\partial x_l} \right) + \delta_{mk} \delta(p,q) = 0, \tag{9}$$

in which $\delta(p,q)$ is the Dirac function with p and q being the source and field points.

Substituting eqn (9) into eqn (8) yields the following integral equation:

$$u_m + \int_\Gamma t^*_{mk} u_k d\Gamma = \int_\Gamma u^*_{mi} t_i \, d\Gamma + \int_\Omega u^*_{mi} b_i d\Omega, \qquad (10)$$

where:

$$t^*_{mk} = D_{ijkl} n_j \frac{\partial u^*_{mi}}{\partial x_l}. \qquad (11)$$

For linear problems, the constitutive tensor D_{ijkl} is constant, therefore an analytical expression for the fundamental solution u^*_{mi} can be solved from eqn (9) [5]. However, for non-linear problems, D_{ijkl} is a function of unknown variables, the displacements or stresses. In this case, eqn (9) is unsolvable for u^*_{mi}. In order to solve such a problem, usually the fundamental solution for corresponding linear problems is employed and, as a result, the additional domain integrals may be generated in the integral eqn (10) [15] and in such cases the body force term b_i in eqn (10) should be understood as a nominal body force.

To evaluate the domain integral involved in eqn (10), the conventional technique is to discretize the domain into internal cells [5], but this eliminates the advantage of BEM that only boundary of the problem needs to be discretized into elements. To overcome this drawback, people usually employ a transformation technique to transfer the domain integral into an equivalent one. The extensively used transformation technique is the dual reciprocity method (DRM) [16]. Another one is the radial integration method (RIM) [17] which can give more accurate results than DRM.

In RIM, the domain integral in eqn (10) can be expressed as:

$$\int_\Omega u^*_{mi} b_i d\Omega = \int_\Gamma \frac{F_{mi}}{r^n(p,q)} \frac{\partial r}{\partial n} d\Gamma(q), \qquad (12)$$

in which the radial integral is:

$$F_{mi} = \int_0^{r(p,q)} u^*_{mi} b_i r^n dr, \qquad (13)$$

where $n = 1$ for 2D and $n = 2$ for 3D problems, respectively. When b_i is a known function, eqns (12) and (13) can give an exact result. However, when b_i includes unknown quantities, such as the non-linear problems, b_i needs to be approximated first by a radial basis function [17] as in DRM, but this may reduce the computational accuracy.

4 LINE-BASED METHODS

The line-based methods include the conventional finite difference method (FDM) [7] and the recently proposed finite line method (FLM) [8]. In these methods, the computational domain is discretized into a series of points and lines to compute the spatial partial derivatives used in the PDEs are formed by around points. FDM constructs the first- and second-order partial derivatives by a line of points along the derivative direction. The main drawback in FDM is that if the lines of defining the derivative directions are not orthogonal each other in 2D or 3D problems, the accuracy of cross-partial derivatives of different directions usually is very poor. This is why FDM cannot well simulate the irregular geometry problems. In contrast, FLM has much better performance in overcoming this drawback.

FLM uses two or three lines (it is called as the line-set) at a collocation point to set up the solution scheme for 2D and 3D problems, as shown in Fig. 5 for a 2D case. Fig. 6 shows the high-order line-sets for an internal collocation point of 2D and 3D problems.

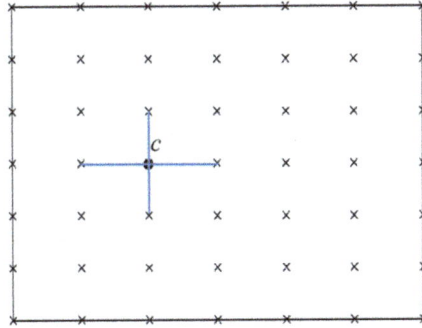

Figure 5: Line-set consisting of two crossed lines for a 2D problem.

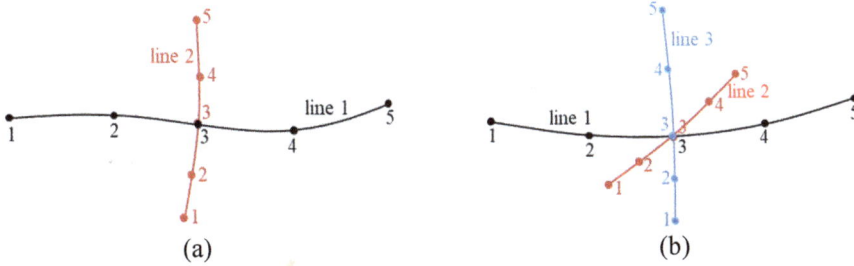

Figure 6:   Node distribution over a line-set defined at a 2D and 3D internal point. (a) 2D problem; and (b) 3D problem.

Along a line of a line-set, the coordinates and physical variable can be expressed as:

$$x_i = \sum_{\alpha=1}^{m} L^\alpha(l)x_i^\alpha \equiv L^\alpha(l)x_i^\alpha, \tag{14}$$

$$u_k = \sum_{\alpha=1}^{m} L^\alpha(l)u_k^\alpha \equiv L^\alpha(l)u_k^\alpha, \tag{15}$$

in which m is the number of nodes defined along a line of the line-set, l is the arclength measured from node 1, and L^α is the Lagrange polynomial:

$$L^\alpha(l) = \prod_{\beta=1, \beta \neq \alpha}^{m} \frac{l - l^\beta}{l^\alpha - l^\beta}, \quad (\alpha = 1 \sim m). \tag{16}$$

By differentiating eqns (14) and (15) with respective to l, we can obtain the expressions of computing the first- and second-order partial derivatives at the collocation point x^c as follows [8]:

$$\frac{\partial u_k(\boldsymbol{x}^c)}{\partial x_i} = d_i^{c\alpha'} u_k^{\alpha'}, \tag{17}$$

$$\frac{\partial^2 u_k(\boldsymbol{x}^c)}{\partial x_i \partial x_j} = d_{ij}^{c\alpha''} u_k^{\alpha''}, \tag{18}$$

where:

$$d_i^{c\alpha'} = \sum_{I=1}^{d} [J]_{iI}^{-1} \frac{\partial L_I^{\alpha'}(l)}{\partial l} \bigg|_{l=l(\boldsymbol{x}^c)}, \tag{19}$$

$$d_{ij}^{c\alpha''} = d_j^{c\beta'} d_i^{\beta'\alpha''}, \tag{20}$$

in which the repeated indexes represent summation, and $d = 2$ for 2D and $d = 3$ for 3D problems, I represents the line number.

Using eqns (17) and (18), we can easily discretize a PDE and related boundary conditions. For example, the PDE for the solid mechanics, shown in eqns (1) and (2) can be written as:

$$D_{ijkl}(\boldsymbol{x}^c) d_{ij}^{\alpha''} u_k^{\alpha''} + d_j^{\beta'} D_{ijkl}^{\beta'}(\boldsymbol{x}^c) d_l^{\alpha'} u_k^{\alpha'} + b_i(\boldsymbol{x}^c) = 0, \quad \boldsymbol{x}^c \in \Omega, \tag{21}$$

$$\begin{cases} u_k(\boldsymbol{x}^c) = \bar{u}_k(\boldsymbol{x}^c), & \boldsymbol{x}^c \in \Gamma_u \\ D_{ijkl}(\boldsymbol{x}^c) n_j(\boldsymbol{x}^c) d_l^{\alpha'} u_k^{\alpha'} = \bar{t}_i(\boldsymbol{x}^c), & \boldsymbol{x}^c \in \Gamma_t. \end{cases} \tag{22}$$

Using above discretized equations, a system of equations can be formed and solved for the displacement u_k at all collocation points.

5 POINT-BASED METHODS

The point-based methods cove a number of numerical methods, such as the mesh free method (MFM) [9], fundamental solution method (FSM) [18], radial basis function method [19] and the newly developed free element method (FrEM) [10]. In these methods, the computational domain is discretized into a series of points, and solution schemes are established by collocating the governing PDEs or its integral forms at each collocation point. In MFM, the partial derivatives at a collocation point c are derived based on a croup of points within a specified support region around c as shown in Fig. 7(a), while in FrEM, partial derivatives are derived based on an isometric free element formed for point c as shown in Fig. 7(b). In

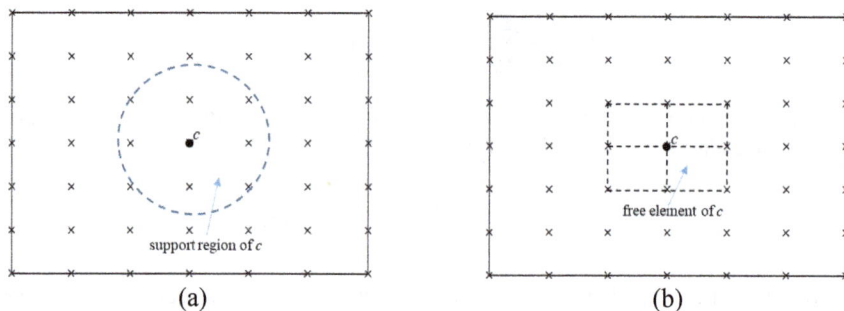

Figure 7: A 2D pattern for MFM (a); and FrEM (b) at a collocation point

MFM and FrEM, both strong-form and weak-form solution schemes are available. In the following, the two schemes of FrEM will be described in detail.

5.1 Strong-form free element method (SFrEM)

In FrEM, a free element as in FEM is independently formed for each collocation point. To easily facilitate forming high-order elements, the Lagrange polynomial element is usually adopted, and its shape functions for high-order dimensions can be straightforwardly formed by the 1D one, as follows [10]:

$$
\begin{aligned}
N_\alpha(\xi,\eta) &= L^I(\xi)L^J(\eta), & \text{for 2D} \\
N_\alpha(\xi,\eta,\zeta) &= L^I(\xi)L^J(\eta)L^K(\zeta), & \text{for 3D,}
\end{aligned}
\tag{23}
$$

where the 1D shape functions L^I, L^J, and L^K are determined by eqn (16), and the superscript α is determined by the permutation of subscripts I, J and K sequentially.

The physical variable and its partial derivatives with respect to global coordinates can be derived as [2], [10]:

$$
u_k = N_\alpha u_k^\alpha,
\tag{24}
$$

$$
\frac{\partial u_k}{\partial x_i} = d_i^{c\alpha} u_k^\alpha, \qquad \frac{\partial^2 u_k}{\partial x_i \partial x_j} = d_{ij}^{c\alpha} u_k^\alpha,
\tag{25}
$$

where:

$$
\begin{aligned}
d_i^{c\alpha} &= \frac{\partial N_\alpha}{\partial x_i} = [J]_{ik}^{-1}\frac{\partial N_\alpha}{\partial \xi_k}, \\
d_{ij}^{c\alpha} &= \frac{\partial^2 N_\alpha}{\partial x_i \partial x_j} = \left[[J]_{ik}^{-1}\frac{\partial^2 N_\alpha}{\partial \xi_k \partial \xi_l} + \frac{\partial [J]_{ik}^{-1}}{\partial \xi_l}\frac{\partial N_\alpha}{\partial \xi_k}\right]\frac{\partial \xi_l}{\partial x_j},
\end{aligned}
\tag{26}
$$

in which $[J]$ is the Jacobian matrix and its inverse and derivatives can be found in Gao et al. [2], [10] for details.

Similar to eqns (21) and (22), the discretized equations for setting up the system of equations can be directly obtained by substituting eqn (25) into eqns (1) and (2).

It is noted that the Lagrange elements shown in eqn (23) have the advantage of easily forming high-order elements. However, for problems with sharply changed geometries where the stress concentration may exist need finer points around these locations. For this type of problems, the weak-form FrEM can give more accurate results even if using coarse elements [20].

5.2 Weak-form free element method (WFrEM)

For a collocation point c with its free element being denoted as Ω_c, let us apply eqn (4) to Ω_c and the weight function is taken as the shape function of the collocation point c in eqn (24), that is $w = N_c$, from which the method is also called as the Galerkin FrEM [20], [21]. Thus, from eqn (4), it follows that:

$$\int_{\Omega_c} \frac{\partial N_c}{\partial x_l} D_{ijkl} \frac{\partial u_k}{\partial x_j} d\Omega = \int_{\partial\Omega_c} N_c t_i(\boldsymbol{x}) d\Gamma + \int_{\Omega_c} N_c b_i d\Omega, \tag{27}$$

in which the derivatives of the shape functions are computed using eqn (26).

Dividing the boundary $\partial\Omega_c$ of the free element Ω_c into two parts, the inner boundary Γ_{Ic} which is located within the problem and out boundary Γ_{oc} which is located on the out surface of the problem, and remembering that N_c is zero on the surfaces not including point c and this makes the integral over Γ_{Ic} zero, eqn (27) can be written as:

$$\int_{\Omega_c} \frac{\partial N_c}{\partial x_l} D_{ijkl} \frac{\partial N_\alpha}{\partial x_j} d\Omega \, u_k^\alpha = \int_{\Gamma_{oc}} N_c \bar{t_i} \, d\Gamma + \int_{\Omega_c} N_c b_i d\Omega, \tag{28}$$

where $\Gamma_{oc} = \partial\Omega_c \cap \partial\Omega$.

From eqn (28) it can be seen that the form of the basic equation in WFrEM is similar to that in the conventional FEM. However, there is an essential difference between them that the element in WFrEM is freely formed at each collocation point, and elements formed for different points are free each other without the restriction of nodal consistency between adjacent elements as in FEM. It is also noted that the elements formed for adjacent collocation points are overlapped in WFrEM, since they are formed locally and independently at each point.

6 NUMERICAL EXAMPLES

In this section, a challenging example with stress concentration is given to compare the described methods. The example is a L-shape plate, as shown in Fig. 8(a). The right side of the plate is fixed and the upper face is applied with a pressure of 1 MPa. The Young's modulus of the plate is 1000 MPa and the Poisson's ratio is 0.3. A uniform mesh of grids is employed as Fig. 8(b) shows.

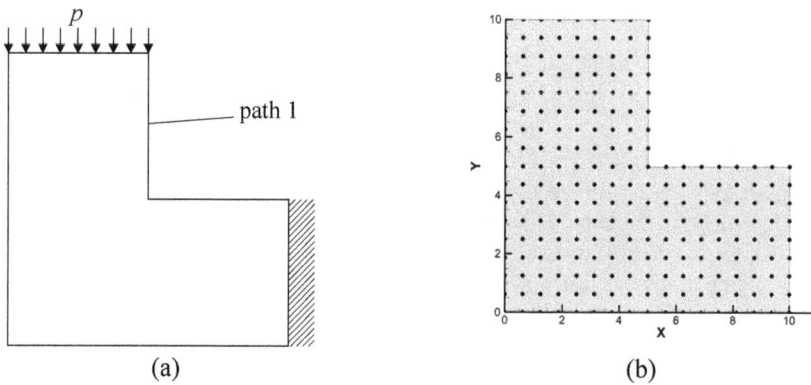

Figure 8: L-shape plate. (a) Geometry; and (b) Grids.

Figs 9 and 10 show the contours of displacement and von Mises stresses computed by FLM. There are three singular points: the two corners on the right side and the concave corner of the plate. Therefore, the stresses are very large at these corners, as shown in Fig. 10. Figs 11 and 12 compare the displacements and von Mises stresses on the bottom edge computed

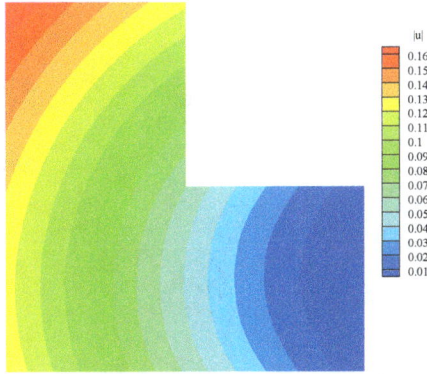

Figure 9: The contours of displacement of the plate.

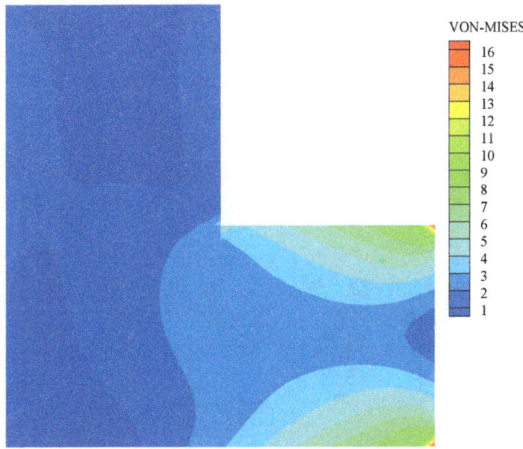

Figure 10: The contours of von Mises stress of the plate.

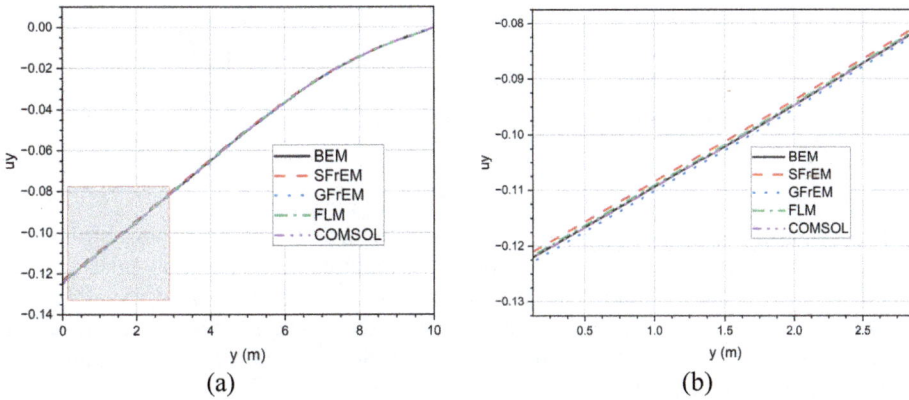

Figure 11: Displacement in y-direction on the bottom edge. (a) Full size; and (b) Local.

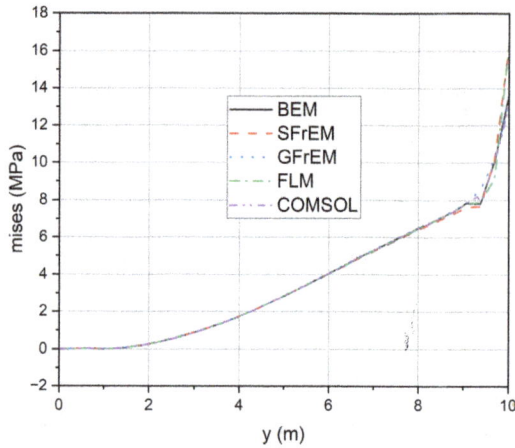

Figure 12: Von Mises stress on the bottom edge.

by different methods. Meanwhile, Figs 13 and 14 compare the displacements and von Mises stresses on path 1. The results of BEM, COMSOL and FLM are almost the same. At the endpoints of the bottom edge and path 1, the displacements of SFrEM are smaller while the displacements of GFrEM are larger than that of BEM and COMSOL. At the singular points, the SFrEM and FLM give larger stresses, which are strong-form schemes.

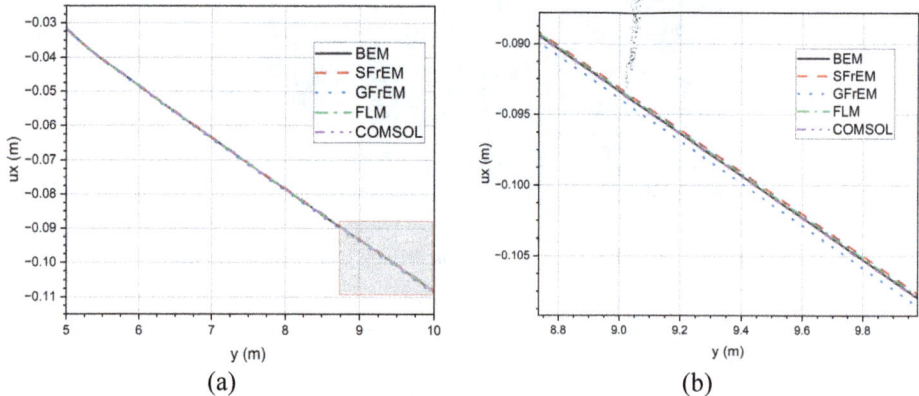

Figure 13: Displacement in y-direction on path 1. (a) Full size; and (b) Local.

7 SUMMARY

In this paper, the progress made in mesh reduction methods of weak-form and strong-form are resumptively described. The focus is on the finite volume method, boundary element method, finite line method and free element method. A numerical example is given for comparison of these methods. The computed results show that all methods can give good results for displacement and the finite line method can simulate the stress concentration more effectively than other method.

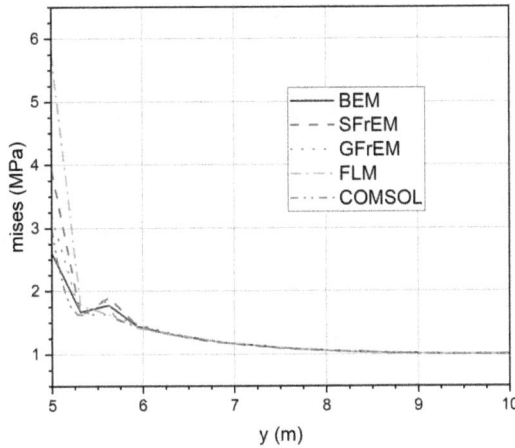

Figure 14: Von Mises stress on path 1.

ACKNOWLEDGEMENT

The support of this investigation by the National Natural Science Foundation of China under Grant No. 12072064 is gratefully acknowledged.

REFERENCES

[1]   Coleman, C.J., On the use of radial basis functions in the solution of elliptic boundary value problems. *Comput. Mech.*, **17**, pp. 418–422, 1996.

[2]   Gao, X.W. et al., Element differential method and its application in thermal-mechanical problems. *International Journal for Numerical Methods in Engineering*, **113**(1), pp. 82–108, 2018.

[3]   Hughes, T.J.R., *The Finite Element Method: Linear Static and Dynamic Finite Element Analysis*, Prentice-Hall: Englewood Cliffs, NJ, 1987.

[4]   Wen, P.H., Cao, P. & Korakianitis, T., Finite block method in elasticity. *Engineering Analysis with Boundary Elements*, **46**, pp. 116–125, 2014.

[5]   Gao, X.W. & Davies, T.G., *Boundary Element Programming in Mechanics*, Cambridge University Press: Cambridge, 2002.

[6]   Onate, E., Cervera, M. & Zienkiewicz, O.C., A finite volume format for structural mechanics. *International Journal for Numerical Methods in Engineering*, **37**(2), pp. 181–201, 1994.

[7]   Wang, H., Dai, W., Nassar, R. & Melnik, R., A finite difference method for studying thermal deformation in a 3D thin film exposed to ultrashort pulsed lasers. *International Journal of Heat and Mass Transfer*, **51**, pp. 2712–2723, 2006.

[8]   Gao, X.-W., Zhu, Y.-M. & Pan, T., Finite line method for solving high-order partial differential equations in science and engineering. *Partial Differential Equations in Applied Mathematics*, 7, 100477, 2023.

[9]   Liu, G.R., An overview on meshfree methods: For computational solid mechanics. *International Journal of Computational Methods*, **13**(5), 1630001, 2016.

[10]   Gao, X.W., Gao, L.F., Zhang, Y., Cui, M. & Lv, J., Free element collocation method: A new method combining advantages of finite element and mesh free methods. *Computers and Structures*, **215**, pp. 10–26, 2019.

[11] Fan, C.M., Huang, Y.K., Chen, C.S. & Kuo, S.R., Localized method of fundamental solutions for solving two-dimensional Laplace and biharmonic equations. *Engineering Analysis with Boundary Elements*, **101**, pp. 188–197, 2019.

[12] Ivankovic, A., Demirdzic, I., Williams, J.G. & Leevers, P.S., Application of the finite volume method to the analysis of dynamic fracture problems. *International Journal of Fracture*, **66**(4), pp. 357–371, 1994.

[13] Gong, J., Xuan, L., Ming, P. & Zhang, W., An unstructured finite-volume method for transient heat conduction analysis of multilayer functionally graded materials with mixed grids. *Numerical Heat Transfer, Part B: Fundamentals*, **63**(3), pp. 222–247, 2013.

[14] Moukalled, F., Mangani, L. & Darwish, M., *The Finite Volume Method in Computational Fluid Dynamics: An Advanced Introduction with OpenFOAM® and Matlab*, Springer, 2016.

[15] Cheng, A.H.D., Chen, C.S., Golberg, M.A. & Rashed, Y.F., BEM for theomoelasticity and elasticity with body force: A revisit. *Engineering Analysis with Boundary Elements*, **25**, pp. 377–387, 2001.

[16] Nardini, D. & Brebbia, C.A., A new approach for free vibration analysis using boundary elements. *Boundary Element Methods in Engineering*, ed. C.A. Brebbia, Springer: Berlin, pp. 312–326, 1982.

[17] Gao, X.-W., The radial integration method for evaluation of domain integrals with boundary-only discretization. *Engineering Analysis with Boundary Elements*, **26**, pp. 905–916, 2002.

[18] Linlin, S., Wen, C. & Cheng, A.H.-D., Method of fundamental solutions without fictitious boundary for plane time harmonic linear elastic and viscoelastic wave problems. *Computers and Structures*, **162**, pp. 80–90, 2015.

[19] Zheng, H., Yang, Z., Zhang, C. & Tyrer, M., A local radial basis function collocation method for band structure computation of phononic crystals with scatterers of arbitrary geometry. *Applied Mathematical Modelling*, **60**, pp. 447–459, 2018.

[20] Xu, B.B. et al., Galerkin free element method and its application in fracture mechanics. *Engineering Fracture Mechanics*, **218**(5), 106575, 2019.

[21] Jiang, W.-W. & Gao, X.-W., Analysis of thermo-electro-mechanical dynamic behavior of piezoelectric structures based on zonal Galerkin free element method. *European Journal of Mechanics: A Solids*, **99**, 104939, 2023.

MESHLESS SOLUTION COMBINED WITH THE DRBEM FOR PLATE BUCKLING ANALYSIS

ALBERTO NUNES RANGEL[1], LEANDRO PALERMO JR[1] & LUIZ CARLOS WROBEL[2,3]
[1]School of Civil Engineering, Architecture and Urban Design, State University of Campinas, Brazil
[2]Department of Mechanical and Aerospace Engineering, Brunel University London, UK
[3]Department of Civil and Environmental Engineering, Pontifical Catholic University of Rio de Janeiro, Brazil

ABSTRACT

The application of the boundary element method to the buckling problem requires a domain integration and the use of the gradient boundary integral equation to perform the analysis. One of the most used techniques to convert the domain integral to equivalent boundary integrals is the dual reciprocity method (DRM). The present study employs the DRM in conjunction with a meshless solution using radial functions to obtain the gradient of the deflection instead of employing the gradient boundary integral equation. The plate bending model considered the effect of shear deformation and the results obtained are compared to those available in the literature.
Keywords: plates (engineering), boundary element method, buckling (mechanics).

1 INTRODUCTION

Plates have been incorporated to the principal structural elements in constructive and mechanical engineering systems, due to the improvement of the constructive processes. This feature has encouraged many researchers to develop theories to describe the bending behaviour of those elements, such as the Kirchhoff and the Reissner–Mindlin models [1], [2].

Although the plate bending behaviour can be appropriately defined with those models, only a few cases can be solved directly. The terms related to the geometrical non-linearity effect (GNL) in the buckling problem modifies the partial differential equations (PDE), and thus it becomes more difficult to obtain the exact solutions.

Some researchers on structural mechanics have developed many studies to analyse the plate buckling behaviour. To improve the accuracy on computation of the buckling parameters for the classical bending model, the minimum potential energy and the Rayleigh–Ritz method were employed by Bryan [3], for uniform compression cases, and by Way [4], for membrane shears combined to bending compression. Similar studies were done by Dawe and Roufaeil [5] and Xiang et al. [6], employing the Mindlin theory. Tham and Szeto [7] used the finite strip method (FSM) to obtain the first and the second buckling modes of square, triangular, parallelogrammical and elliptical geometries. The finite element method (FEM) has been widely applied to plate buckling, like in Sakiyama and Matsuda [8] and Featherson and Ruiz [9]. The boundary element method (BEM) was used in the buckling analyses considering the classical bending model [10] or those with the effect of shear deformation in bending [11]. The boundary discretization is the main advantage of the BEM formulation, a small system of equations related to the boundary node values results in comparison to that in the domain discretization methods. The internal values required for the buckling analysis are subsequently computed after the solution of the system of equations. Nardini and Brebbia proposed the dual reciprocity method (DRM) [14] to keep the advantage of boundary only discretization in problems requiring the domain integral for the solution. The boundary integral equations of the problem are combined with the particular solutions used to convert the domain integrals in a sum of boundary integrals. The generalization of the DRM for several cases found in engineering problems as well as a complete explanation on the technique was done in Partridge et al. [15].

WIT Transactions on Engineering Sciences, Vol 135, © 2023 WIT Press
www.witpress.com, ISSN 1743-3533 (on-line)
doi:10.2495/BE460101

The dual reciprocity boundary element method (DRBEM) was used in problems containing the GNL effect, to compute the plate buckling parameters for rectangular and circular geometries [16], and to analyse large deflections of plates in Purbolaksono and Aliabadi [17], all of them based on the Reissner–Mindlin model. Considering the Kirchhoff's model, the DRBEM was combined with the analog equations method (AEM) in Chinnaboon et al. [18], to get the bucking factors of rectangular plates and in Katsikadelis and Babouskos [19] for post-buckling analysis. More complex applications using the multi-region method were done in Baiz and Aliabadi [20], on the local buckling of thin-walled structures using the classical model, on large deflections of assembled structures with shear deformable plates and in fracture mechanics including the GNL effect [22].

The present study is derived from Soares et al. [23] where the formulation for plate buckling considered the effect of shear deformation. No relation was required for in-plane stress derivatives and the GNL effect used the gradient of the deflection instead of the second derivatives of the deflection, requiring the null value for the divergence of the in-plane stress tensor, which is employed in most of the BEM formulations. The use of a meshless solution for the gradient of deflection instead of using the gradient boundary integral equation is the main difference in the present study with reference to Soares et al. [23]. The buckling parameter of unilaterally compressed square plates were obtained considering the simply supported and the clamped edge conditions for cases with thickness to plate side ratio between 0.001 to 0.2.

2 BOUNDARY INTEGRAL EQUATIONS

The linearization of the geometrical non-linearity (GNL) effect in the plate bending equilibrium under the in-plane forces on the domain results in the plate buckling equations [1]. The BEM formulation for the buckling analysis in isotropic plates [11] employed the displacement and the deflection gradient BIEs. The equations are next presented with the values {1, 2 and 3} for the Latin indices and {1, 2} for the Greek indices. The differentiation in the kernels of the gradient BIE was written in terms of the field point coordinates:

$$
C_{ij}(x')u_j(x') + \int_\Gamma \left[T_{ij}(x',x)u_j(x) - U_{ij}(x',x)t_j(x)\right]d\Gamma(x) = \cdots
$$

$$
= \int_\Gamma n_\alpha(x)N_{\alpha\beta}(x)u_{3,\beta}(x)U_{i3}(x',x)d\Gamma(x) - \iint_\Omega N_{\alpha\beta}(X)u_{3,\beta}(X)U_{i3,\alpha}(x',X)d\Omega(X) \quad (1)
$$

$$
u_{3,\gamma}(X') = \int_\Gamma \left[T_{3j,\gamma}(X',x)u_j(x) - U_{3j,\gamma}(X',x)t_j(x)\right]d\Gamma(x) +
$$

$$
\cdots - \int_\Gamma n_\alpha(x)N_{\alpha\beta}(x)u_{3,\beta}(x)U_{i3,\gamma}(X',x)d\Gamma(x) + \cdots
$$

$$
\cdots + \iint_\Omega N_{\alpha\beta}(X)u_{3,\beta}(X)U_{33,\alpha\gamma}(X',X)d\Omega(X) \quad (2)
$$

where C_{ij} is an element of the matrix C related to the boundary geometry at the source point, being 0,5 for the external points and 1,0 for internal points, which becomes the identity matrix when a smooth boundary is considered, u_α is the plate rotation in direction α, and u_3 is the plate deflection. U_{ij} represents the rotation ($j = 1, 2$) or the deflection ($j = 3$) due to a unit couple ($i = 1, 2$) or a unit point force ($i = 3$). T_{ij} represents the moment ($j = 1, 2$) or the shear

($j = 3$) due to a unit couple ($i = 1, 2$) or a unit point force ($i = 3$). U_{ij} and T_{ij} are related to the fundamental solution. $N_{\alpha\beta}$ is the resultant on the thickness of the in-plane stresses.

The constitutive equations for the isotropic plates are given by:

$$M_{\alpha\beta} = D \frac{(1-v)}{2} \left(u_{\alpha,\beta} + u_{\beta,\alpha} + \frac{2v}{1-v} u_{\gamma,\gamma} \delta_{\alpha\beta} \right) \tag{3}$$

$$Q_\alpha = D \frac{(1-v)}{2} \lambda^2 \left(u_\alpha + u_{3,\alpha} \right) \tag{4}$$

with

$$\lambda^2 = 12 \frac{\kappa^2}{h^2}$$

D is the flexural rigidity, h is the plate thickness, v is Poisson's ratio, and $\delta_{\alpha\beta}$ is the Kronecker delta. The shear parameter κ^2 is equal to 5/6 or $\pi^2/12$ for the Reissner or the Mindlin model, respectively.

3 APPLICATION OF THE DUAL RECIPROCITY METHOD

The kernels in the domain integrals of eqns (1) and (2) contain the products between the GNL effect (plate deflection weighted by in-plane forces) and the first and the second derivatives of deflection due to the fundamental solution (U_{i3}), respectively. The introduction of the DRM in eqns (1) and (2) requires the use of the first derivative of the BIEs for the displacements (rotation and deflection) as well as the second derivatives of the BIE for the deflection [23].

$$u_{i,\gamma}(X') = \int_\Gamma \left[T_{ij,\gamma}(X',x)u_j(x) - U_{ij,\gamma}(X',x)t_j(x) \right] d\Gamma(x) - \iint_\Omega q(X)U_{i3,\gamma}(X',X)d\Omega(X) \tag{5}$$

$$u_{3,\alpha\gamma}(X') = \int_\Gamma \left[T_{ij,\alpha\gamma}(X',x)u_j(x) - U_{ij,\alpha\gamma}(X',x)t_j(x) \right] d\Gamma(x) -$$
$$\iint_\Omega q(X)U_{i3,\alpha\gamma}(X',X)d\Omega(X) \tag{6}$$

The order of the singularities in eqns (1) and (2) are increased in the DRM with the introduction of eqns (5) and (6). The meshless formulation to obtain the gradient of the deflection carried to only use the eqn (3) in the DRM. In this way, the singularities remain in the same order of those in the formulation [11] with eqns (1) and (2). The gradient BIE (eqn 5) is introduced in eqn (3) to work with the DRM, which is combined with the meshless formulation.

The steps to introduce the DRM were the same in Soares et al. [23] but they were only applied to eqn (3). A vector function (b) related to the GNL effect and resulting from the product between the in-plane force tensor and the gradient of plate deflection is given by:

$$b_\theta(X) = N_{\theta\beta}(X)u_{3,\beta}(X) \tag{7}$$

The vector function (b) is approximated with the following relation [15]:

$$b_\theta(X) = \sum_{m=1}^{N+L} \alpha_\theta^m f^m \tag{8}$$

The summation in eqn (8) is extended to all points employed in the DRM, i.e., the total number of points placed on the boundary (N) and in the domain (L), f^m and α_θ^m are sets of the approximating functions and weighting coefficients, respectively [15]. The application of the DRM employs the particular solution $\hat{u}_j^m$, which is related to the approximating function f^m [23].

The radial basis function $(1 + r)$, which was used in Soares et al. [23], carried the following particular solutions:

$$\hat{u}_\alpha^m = \frac{1}{D}\left(\frac{r^3}{16} + \frac{r^4}{45}\right)r_{,\alpha}\ ; \quad \hat{u}_3^m = \frac{2}{D(1-v)\lambda^2}\left(\frac{r^2}{4} + \frac{r^3}{9}\right) - \frac{1}{D}\left(\frac{r^4}{64} + \frac{r^5}{225}\right) \tag{9}$$

The distributed shear and bending moments related to the particular solutions are obtained with the constitutive eqns (3) and (4), and reduced to the generalized tractions with the following relations:

$$\hat{t}_\alpha^m = \hat{M}_{\alpha\beta}^m n_\beta$$
$$\hat{t}_3^m = \hat{Q}_\beta^m n_\beta \tag{10}$$

The displacement BIE for the buckling problem using the DRM is obtained after the introduction of the relation given in eqn (8) in the kernel of the domain integral in eqn (3) and using the gradient BIE, eqn (5). The final displacement BIE is given by Soares et al. [23]:

$$C_{ij}(x')u_j(x') + \int_\Gamma \left[T_{ij}(x',x)u_j(x) - U_{ij}(x',x)t_j(x)\right]d\Gamma(x) = \cdots$$

$$= \int_\Gamma n_\alpha(x)N_{\alpha\beta}(x)u_{3,\beta}(x)U_{i3}(x',x)d\Gamma(x) + \sum_{m=1}^{N+L} \alpha_\theta^m \left\{c\hat{u}_{i,\theta}^m(x') + \cdots\right.$$

$$\cdots\cdots - \int_\Gamma \left[n_\alpha(x)M_{i\alpha\beta,\theta}(x',x)\hat{u}_\beta^m(x) + n_\beta(x)Q_{i\beta,\theta}(x',x)\hat{u}_3^m(x) + \cdots\right.$$

$$\cdots - U_{i\beta,\theta}(x',x)\hat{t}_\beta^m(x) - U_{i3,\theta}(x',x)\hat{t}_3^m(x)\right]d\Gamma(x)\bigg\} \tag{11}$$

The gradient of the deflection for the buckling analysis was obtained with the meshless formulation [15]. The meshless solution employed a different radial basis function to that in eqn (8) for the DRM. The relation between the deflection and the radial functions was established in the same way of eqn (8) and is given by:

$$u_3(X) = \sum_{m=1}^{N+L} \beta^m f_w^m \tag{12}$$

The gradient of the deflection considering eqn (12) is next written with the differentiation done on the radial basis function, i.e.:

$$u_{3,\theta}(X) = \sum_{m=1}^{N+L} \beta^m f_{w,\theta}^m \tag{13}$$

The gradient of the deflection is obtained when eqns (12) and (13) are combined, which can be written in the matrix form:

$$(u_3) = [f_w](\beta) \Rightarrow (\beta) = [f_w]^{-1}(u_3)$$
$$(u_{3,\theta}) = [f_{w,\theta}](\beta)$$
$$(u_{3,\theta}) = [f_{w,\theta}][f_w]^{-1}(u_3) \tag{14}$$

Two types of augmented thin plate spline RBF were employed in the numerical examples and are given by:

$$f_w^m = \left(\frac{r}{c}\right)^4 \ln\left[1 + \frac{r}{c} + \left(\frac{r}{c}\right)^2 + \left(\frac{r}{c}\right)^3\right] \tag{15}$$

$$f_w^m = \left(\frac{r}{c}\right)^3 \ln\left[1 + \frac{r}{c} + \left(\frac{r}{c}\right)^3 + \left(\frac{r}{c}\right)^5\right] \tag{16}$$

4 NUMERICAL IMPLEMENTATION

The discretization of the BIEs employed quadratic isoparametric boundary elements with the collocation points always placed on the boundary. The same mapping function was used for conformal and non-conformal interpolations. The collocation points were placed at positions (-0.67 and 0.0), in the range (-1, $+1$), in the case of continuous elements and at positions (-0.67, 0.0, $+0.67$) in the case of discontinuous elements, i.e., the collocation points were always shifted inside the boundary elements. The singularity subtraction [13] and the transformation of variable techniques [13] were employed for the Cauchy and weak-type singularities, respectively, when integrations were performed on elements containing the collocation points. The standard Gauss–Legendre scheme was employed for integrations on elements not containing the collocation points.

The DRM considered points distributed in the domain and on the boundary. The points on the boundary were placed at the positions of the collocation points. The first boundary integral on the right-hand side (RHS) of eqn (11) is not related with the DRM but the result of the algebraic manipulation on the integral containing the GNL effect in Soares and Palermo [11]. The discretization of the DBIE in eqns (1) or (11) assumed a constant value along each boundary element for the sum of products between the gradient of the deflection and the in-plane forces, which values were obtained at the central node of the boundary elements. It is necessary to note the value the deflection gradient was computed with eqn (14) in this study.

The eigenvalue analysis used the basic inverse iteration with the Rayleigh quotient [23] according to the following equations:

$$Ax^{(k+1)} = Bx^{(k)} \tag{17}$$

$$\lambda_k = \frac{(x^{(k+1)}, x^{(k)})}{(x^{(k+1)}, x^{(k+1)})} \tag{18}$$

The vector $x^{(k)}$ in eqns (17) and (18) is related to values of the gradient of the deflection at the DRM points. Starting with an eigenvector $x^{(0)}$ with elements equal to 1.0, the values of the displacements and tractions at the boundary nodes are found with eqn (11), and the deflection values are introduced in eqn (14) to get the gradients, which were used to obtain the eigenvector $x^{(1)}$. The vectors $x^{(0)}$ and $x^{(1)}$ are introduced in eqn (18) to obtain the eigenvalues.

5 NUMERICAL EXAMPLES

A square plate under uniform load on the domain was solved and the results obtained with the meshless solution (eqn 14) were compared with the BEM solution (eqn 5) [12]. The Young's modulus (E) and Poisson's ratio (ν) were 2 MPa and 0.3, respectively. The shear parameter κ^2 was $\pi^2/12$ (Mindlin), the plate side (L) and thickness (h) were 4 m and 0.8 m, respectively, and the uniform load q was 0.64 kN.m^{-2}. The boundary conditions were the clamped edge (C), the free edge (F) and the simply supported edge (S) with the hard condition (the twist rotation restrained). The meshes used 128 quadratic boundary elements (BE) combined with 49 internal points (P) uniformly distributed in the domain, 121 (P) and 225 (P). The results obtained with 49 internal points are presented because better values were obtained when the number of internal points was increased. The radial function for the meshless solution was presented in eqn (15). The relative differences between the obtained results are presented in Tables 1 and 2. The relative differences were increased in cases with free edges.

WIT Transactions on Engineering Sciences, Vol 135, © 2023 WIT Press
www.witpress.com, ISSN 1743-3533 (on-line)

Table 1: Relative differences between results obtained for deflection gradient with the BEM solution (eqn 5) and the meshless solution (eqn 14) for plates with same boundary conditions on opposite sides.

Position		Relative differences (%)									
x_1/L	x_2/L	$u_{3,1}$	$u_{3,2}$	$u_{3,1}$	$u_{3,2}$	$u_{3,1}$	$u_{3,2}$	$u_{3,1}$	$u_{3,2}$	$u_{3,1}$	$u_{3,2}$
0.125	0.125	0.8	0.8	1.1	1.1	2.3	1.6	9.3	0.3	14.7	1.6
0.250	0.125	0.4	0.2	0.6	2.0	1.4	2.7	6.7	0.3	1.5	0.6
0.375	0.125	0.3	0.0	0.4	2.3	1.1	3.1	5.7	0.2	1.3	0.3
0.500	0.125	–	0.0	–	2.4	–	3.1	–	0.2	–	0.2
0.125	0.250	0.2	0.4	2.0	0.6	1.1	0.8	6.7	0.1	30.6	0.6
0.250	0.250	0.1	0.1	0.9	0.9	0.8	1.2	5.1	0.1	19.5	0.3
0.375	0.250	0.1	0.0	0.7	1.1	0.6	1.4	4.5	0.1	16.5	0.1
0.500	0.250	–	0.0	–	1.1	–	1.4	–	0.1	–	0.1
0.125	0.375	0.0	0.3	2.3	0.4	0.8	0.6	5.5	0.1	25.4	0.3
0.250	0.375	0.0	0.1	1.1	0.7	0.5	0.9	4.1	0.1	21.2	0.2
0.375	0.375	0.0	0.0	0.7	0.7	0.4	0.9	3.6	0.1	19.6	0.1
0.500	0.375	–	0.0	–	0.8	–	1.0	–	0.1	–	0.1
0.125	0.500	0.0	–	2.4	–	0.7	–	5.2	–	23.1	–
0.250	0.500	0.0	–	1.1	–	0.4	–	3.9	–	20.0	–
0.375	0.500	0.0	–	0.8	–	0.3	–	3.4	–	19.1	–
0.500	0.500	–	–	–	–	–	–	–	–	–	–
Boundary conditions		SSSS		CCCC		CSCS		SFSF		CFCF	

Table 2: Relative differences between results obtained for deflection gradient with the BEM solution (eqn 5) and the meshless solution (eqn 14) for plates with same boundary conditions on opposite sides in one direction only.

Position		Relative differences (%)			
x_1/L	x_2/L	$u_{3,1}$	$u_{3,2}$	$u_{3,1}$	$u_{3,2}$
0.125	0.125	15.1	1.9	0.7	2.4
0.375	0.125	24.8	0.1	0.3	2.8
0.625	0.125	25.2	0.1	0.3	2.8
0.875	0.125	15.3	1.9	0.7	2.4
0.125	0.375	2.3	0.1	0.2	0.3
0.375	0.375	1.9	0.0	0.1	0.4
0.625	0.375	1.5	0.0	0.1	0.4
0.875	0.375	1.9	0.1	0.2	0.3
0.125	0.625	4.7	0.0	0.0	0.7
0.375	0.625	2.4	0.0	0.0	0.4
0.625	0.625	2.3	0.0	0.0	0.4
0.875	0.625	4.5	0.0	0.0	0.6
0.125	0.875	4.4	0.2	0.2	4.3
0.375	0.875	1.6	0.1	0.0	2.9
0.625	0.875	1.6	0.1	0.0	2.9
0.875	0.875	4.4	0.2	0.2	4.2
Boundary conditions		CFFF		CSFS	

The buckling parameters were obtained for a square plate under uniaxial compression condition shown in Fig. 1, considering the simply supported and clamped edge conditions, and using 128 quadratic boundary elements and 676 internal points. The results obtained for those cases are presented in Table 3, and were very close to the literature. They are shown to evaluate the application of the meshless solution combined with the DRM for a non-linear problem. The buckling parameter k is a non-dimensional value related to the critical load of the plate (N_{cr}), the length of the plate side (a) and the flexural rigidity (D) according to following expression:

$$k = \frac{a^2 N_{cr}}{\pi^2 D} \tag{19}$$

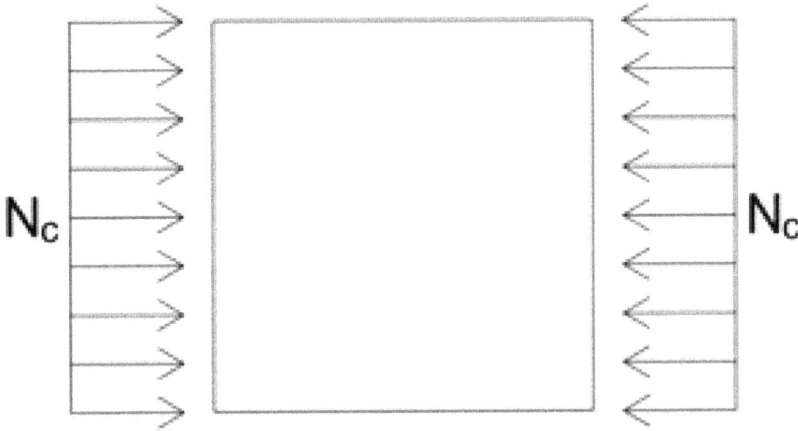

Figure 1: Square plate under uniaxial compression.

The plates had a thickness to length ratio between 1/1000 and 1/5. The Young's modulus (E) and Poisson's ratio (ν) were 206.9 GPa and 0.3, respectively. The value of the shear parameter κ^2 was 5/6. The comparison between values for the buckling parameter obtained with the results available literature are presented in Table 1.

6 CONCLUSIONS

The present technique combining the DRM to convert the domain integral to equivalent boundary integrals and the meshless solution to compute the deflection derivatives instead of using the BIE for the gradient of the defection was efficient to compute the buckling parameters. The displacement BIE containing terms resulting from the DRM was the only equation in the numerical formulation. The equation was used to obtain the boundary element matrices as in all boundary element formulations and to compute the required deflection at the points needed to obtain the gradient of the deflection, which is used to introduce the GNL effect. The main feature with reference to the formulation presented in Soares et al. [23] was the use of one equation type (displacement BIE) in the solution instead of working with the gradient BIE containing integrals with the singularities increased. The computer and/or numerical effort in this technique was low whereas the choice of the better radial function to perform an efficient analysis was the main difficulty. The authors intend to apply the present technique to all cases studied in Soares and Palermo [11] and Soares et al. [23] and relevant cases in the literature.

Table 3: Buckling parameters for the square plate obtained with the DRM combined with the meshless solution for the deflection derivatives.

Type	h/a	[21]	Obtained	Difference (%)
1. SSSS	0.001	4.0000*	4.0931	2.328
	0.01	3.9977	3.9972	0.013
	0.05	3.9444*	3.9448	0.010
	0.100	3.7865*	3.7868	0.008
	0.200	3.2637*	3.2639	0.006
2. CCCC	0.01	10.1382*	10.0464	0.905
	0.05	9.5588*	9.5666	0.082
	0.1	8.2917*	8.2993	0.092
	0.2	5.3156*	5.3448	0.549
3. CSSS	0.001	5.7401	5.7516	0.200
	0.05	5.5977	5.6003	0.046
	0.100	5.2171	5.2238	0.128
	0.200	4.1364	4.1514	0.363
4. SSSC	0.001	4.8471	4.9969	2.013
	0.05	4.7454	4.7495	0.086
	0.1	4.4656	4.4724	0.152
	0.2	3.6115	3.6254	0.385
5. CSCS	0.001	7.6911	7.6880	0.040
	0.05	7.2989	7.3109	0.164
	0.1	6.3698	6.3896	0.311
	0.2	4.3204	4.3480	0.639
6. CSCS	0.001	6.7431	6.5317	3.135
	0.05	6.5238	6.5316	0.120
	0.1	5.9487	5.9626	0.234
	0.2	4.4004	4.4243	0.543

*[23]

REFERENCES

[1] Timoshenko, S. & Woinowsky-Krieger, S., *Theory of Plates and Shells*, McGraw-Hill: New York, 1959.

[2] Timoshenko, S. & Gere. J., *Theory of Elastic Stability*, New York: McGraw-Hill, vol. 294, 1961.

[3] Bryan, G.H.. On the Stability of a plane plate under thrusts in its own plane, with applications to the "buckling" of the sides of ship. *Proceedings of the London Mathematical Society*, 1(1), pp. 54–67, 1890.

[4] Way, S., Stability of rectangular plates under shear and bending forces. *Journal of Applied Mechanics*, 16, 1936.

[5] Dawe, D.J. & Roufaeil, O.L., Buckling of rectangular Mindlin plates. *Computers and Structures*, 15(4), pp. 461–471, 1982.

[6] Xiang, Y. et al., Mindlin plate buckling with prebuckling in-plane deformation. *Journal of Engineering Mechanics*, vol. 119, n. 1. pp. 1–18, 1993.

[7] Tham, L. & Szeto, H., Buckling analysis of arbitrarily shaped plates by spline finite strip method. *Computers and Structures*, 36(4), pp. 729–735, 1990.

[8] Sakiyama, T. & Matsuda, H., Elastic buckling of rectangular Mindlin plate with mixed boundary conditions. *Computers and Structures*, **25**(5), pp. 801–808, 1987.

[9] Featherson, C.A. & Ruiz, C., Buckling of flat plates under bending and shear. Department of Engineering Science, University of Oxford, 1998.

[10] Bezine, G., Cimetiere, A. & Gelbert, J.P., Unilateral buckling of thin elastic plates by the boundary integral equation method. *International Journal for Numerical Methods in Engineering*, **21**(12), pp. 2189–2199, 1985.

[11] Soares Jr., R.A. & Palermo Jr., L., Effect of shear deformation on the buckling parameter of perforated and non-perforated plates studied using the boundary element method. *Engineering Analysis with Boundary Elements*, **85**, pp. 57–69, 2017.

[12] Weeën, F.V., Application of the boundary integral equation method to Reissner's plate model. *International Journal for Numerical Methods in Engineering*, **18**(1). pp. 1–10, 1982.

[13] Wrobel, L.C., *The Boundary Element Method: Applications in Thermo-Fluids and Acoustics*. John Wiley: Chichester, UK, 2002.

[14] Nardini, D. & Brebbia, C.A., A new approach to free vibration analysis using boundary elements. *Applied Mathematical Modelling*, **7**(3), pp. 157–162, 1983.

[15] Partridge, P.W., Brebbia, C.A. & Wrobel, L.C., *The Dual Reciprocity Boundary Element Method*, Computational Mechanics Publication and Elsevier Applied Science: London and New York, 1992.

[16] Purbolaksono, J. & Aliabadi, M.H., Buckling analysis of shear deformable plates by boundary element method. *International Journal for Numerical Methods in Engineering*, **62**, pp. 537–563, 2004.

[17] Purbolaksono, J. & Aliabadi, M.H., Large deformation of shear deformable plate by boundary element method. *Journal of Engineering Mathematics*, **51**, pp. 211–230, 2005.

[18] Chinnaboon, B., Chucheepsakul, S. & Katsikadelis, J.T., A BEM-based meshless method for elastic buckling analysis of plates. *International Journal of Structural Stability and Dynamics*, **7**(1). pp. 81–99, 2007.

[19] Katsikadelis, J.T. & Babouskos, N., The post-buckling analysis of plates: A BEM based meshless variational solution. *Automatic Control and Robotics*, **6**, pp. 113–118, 2007.

[20] Baiz, P.M. & Aliabadi, M.H., Local buckling of thin-walled structures by the boundary element method. *Engineering Analysis with Boundary Elements*, **33**(3). pp. 302–313, 2009.

[21] Hosseini-Hashemi, S., Khorshidi, K. & Amabili, M., Exact solution for linear buckling of rectangular Mindlin plates. *Journal of Sound and Vibration*, **315**(1), pp. 318–342, 2008.

[22] Purbolaksono, J., Dirgantara, T. & Aliabadi, M.H., Fracture mechanics analysis of geometrically nonlinear shear deformable plates. *Engineering Analysis with Boundary Elements*, **36**, pp. 87–92, 2012.

[23] Soares Jr., R.A., Palermo Jr., L. & Wrobel, L.C., Application of the dual reciprocity method for the buckling analysis of plates with shear deformation. *Engineering Analysis with Boundary Elements*, **106**, pp. 427–439, 2019.

SECTION 5
FLUID FLOW MODELLING

NOVEL BOUNDARY INTEGRAL METHOD FOR SLOW FREE SURFACE FLOWS

LOÏC GOBET & ROBERT G. OWENS*
Département de Mathématiques et de Statistique, Université de Montréal, Canada

ABSTRACT

The present article introduces a novel boundary integral method (BIM), adapted from an earlier method of Hansen and Kelmanson (1992, 1994) and suitable for the solution of creeping flow boundary value problems where the boundary presents singularities in the stresses. We use the new BIM to solve the problem of the planar extrusion of a Newtonian fluid at zero Reynolds number and, in particular, to determine the shape of the free surface in the immediate neighbourhood of the separation point for a range of capillary numbers. The proposed method incorporates the singular solution near the separation point, thus overcoming one limitation of a classical BIM to the problem (see, for example, Kelmanson (1983)). In a recent article, Owens (2022) also incorporated the singular solution into his BIM formulation. However, since the integration path used in the present BIM passes directly through the separation point this leads to an important improvement on the method of Owens (2022), who was obliged to skirt the singularity due to the non-integrability there of the normal derivative of the vorticity. Results presented for the extrudate swell ratio, the angle of separation and the leading exponent in the asymptotic expression for the stream function are shown to be in convincing agreement with others in the theoretical, numerical and experimental literature.
Keywords: Stokes flow, singularity, boundary integral method, free surface flows.

1 INTRODUCTION

The use of boundary integral methods (BIMs) for the solution of Poisson's equation and the biharmonic equation arising in boundary value problems in such fields as linear elasticity, electrostatics, potential flows and slow viscous flows dates back some fifty years. That such methods only require boundary data for the computation of the solution throughout the domain of definition presents both advantages and disadvantages. On the positive side, and where applicable, BIMs are, in general, more cost-efficient than other mesh-based methods, giving rise to much smaller linear systems to be solved than those arising from equivalent finite element or finite difference (FD) approximations, for example, albeit with matrices that may be ill-conditioned. However, the presence of singularities on the boundary due to a sharp corner or a sudden change of boundary condition, for example, leads to poor convergence of classical BIMs and the need to treat such singularities in a mathematically rigorous fashion has spawned a number of different techniques in the literature.

1.1 BIMs for boundary value problems with singularities on the boundary

In 1983, Kelmanson [1], [2] used a singular subtraction (SS) BIM to solve three problems: the steady lid-driven cavity problem, the problem of the slow channel flow of a viscous fluid through a sudden expansion and the planar "stick-slip" problem for a viscous fluid. The dominant asymptotic behaviour of the stream function ψ in the neighbourhood of the singularities in these problems was written down by Kelmanson in the form of truncated series g whose coefficients had to be determined as part of the solution process. Provided that

ORCID: http://orcid.org/0000-0002-9907-2586

WIT Transactions on Engineering Sciences, Vol 135, © 2023 WIT Press
www.witpress.com, ISSN 1743-3533 (on-line)
doi:10.2495/BE460111

the truncated series had a sufficient number of terms, the recasting of the problems in terms of a modified stream function

$$\chi := \psi - g, \tag{1}$$

meant that the modified flow variables became regular throughout the flow domains. By evaluating the kernel integrals exactly rather than by Gaussian quadrature computer time savings of up to 38% were realised and, in the first of the two articles, CPU time and storage savings compared with FD methods were shown to be substantial.

The subtraction of a truncated series representation of the asymptotic behaviour at a boundary singularity of the solution to an elliptic boundary value problem has been done for boundary value problems involving harmonic functions and this dates back to the SS method of Symm [3]. Ingham and Kelmanson [4] used both a classical BIM and a BIM modified using the SS method to solve two Laplacian boundary value problems for a potential function φ. One problem (A) had Dirichlet conditions prescribed for φ where the singularity was due to a discontinuous boundary potential and the other problem (B) featured a discontinuity in the prescribed boundary flux. A comparison with FD calculations, where the finite difference boundary mesh points coincided with the BIM segment end points, showed that for finer meshes the two BIMs used considerably less CPU time for problem B. In the case of problem A, it was only the possibility of calculating an optimal relaxation parameter for the FD SOR iterations that allowed that method to require less CPU time to converge than the BIMs. The modified BIM was considerably more accurate than the classical BIM for problem A, and only required about 10% more CPU time.

In 1984, Kelmanson [5] and Ingham and Kelmanson [6] used a singularity incorporation (SI) BIM to solve nonlinear problems of two dimensional steady state heat transfer, the conducting medium having variable thermal conductivity. Under a Kirchoff transformation the transformed variable T satisfied Laplace's equation and the series representation of the behaviour of T in the neighbourhood of a boundary singularity was enforced solely on a small number of elements nearest to the singularity, in contrast to the SS method of Symm [3]. Thus the modification to the classical BIM was small although the authors reported dramatic improvements in the rate of convergence of results throughout the solution domain of the problems considered.

Where applicable, the so-called singularity annihilation (SA) BIM, employed by Kelmanson and Lonsdale [7], has the advantage over both SS and SI BIMs in that it requires no extra coding and employs the natural boundary conditions of the original problem. The basic idea is a simple one: if the boundary singularities are in a region of the boundary where suitably constructed Green's functions are asymptotically small then the solution's singular behaviour may be annihilated. This was achieved by the authors for the benchmark lid-driven cavity problem where the asymptotic behaviour $\psi = \mathcal{O}(r)$ of the stream function ψ in corners between the moving lid and the vertical walls is the same as that of Taylor's scraper flow problem [8] and leads to $\mathcal{O}(r^{-1})$ behaviour of the vorticity and pressure at these points: a particularly severe test of the numerical method. Comparisons of the calculated stream function values with those obtained using the SS method of Kelmanson [1] showed impressive agreement and were corroborated by the biharmonic method of fundamental solutions of Karageorghis and Fairweather [9]. However, the SA BIM required substantially less coding than the SS BIM of Kelmanson [1].

The $\mathcal{O}(r^{-1})$ behaviour of the stresses at the singular points between the lid and walls in the lid-driven cavity problem mean that they are not integrable and that therefore it is not a physically realizable flow. Hansen and Kelmanson [10], [11] used a BIM to investigate the effect of introducing small leaks at the corners where the lid meets the walls, thus making the flow physical and modelling what must take place under experimental conditions. Near the re-entrant corners in the leaky lid-driven cavity problem the authors showed that the stream function $\psi = \mathcal{O}(r^{1+\lambda})$ with λ real and in the interval $(0, 1)$. Thus, the vorticity ω and pressure p were now integrable but the normal derivative of the vorticity $\partial \omega / \partial n$, appearing in the integral formulae for ψ and ω, was not. However, the authors were able to evaluate these integrals without avoiding the re-entrant corner by noting that on a piecewise smooth boundary in Stokes flow the tangential derivative of the pressure equals $\partial \omega / \partial n$ (where these exist) and that with a change of variables and integration by parts the integrals in the integral formulae were finite. The resulting BIM was first used by Xu [12].

Some 65 years ago, Michael [13] showed for the problem of creeping planar flow of a Newtonian fluid extruding from a channel that, with the assumption of either vanishing surface tension or zero curvature of the free surface near the separation point, the angle formed between the free surface and the channel wall was $180°$. Since that time there has been widespread disagreement, both in the experimental and numerical literature, about the character of the free surface (and therefore about the solution to the problem) in the neighbourhood of the channel orifice and, in particular, about the angle of separation. The values of the separation angles computed by Owens [14] for capillary numbers Ca ranging from 1 to 1,000 fell well within the interval of values published in previous experimental and numerical papers, although there is a paucity of such results available in the literature. His analysis showed that for non-zero surface tensions the normal stress and curvature of the free surface were unbounded at the point of separation, consistent with the analysis and results of Anderson and Davis [15], Schultz and Gervasio [16] and Salamon et al. [17]. It was concluded by Owens that the case of zero surface tension ($Ca = \infty$, with corresponding separation angle of $180°$), is a singular limit.

The problem of the planar extrusion of a Newtonian fluid at vanishing Reynolds number Re was considered by Kelmanson [18] using a classical BIM but no account of the singularity in the stresses at the separation point C was taken, leading to the author's admission that it was to be expected that near this point the numerical results would be in error. Owens [14] has also used a BIM to solve the same problem, with the incorporation of the singular solution near the separation point. However, due to the non-integrability at C of $\partial \omega / \partial n$, appearing in the integral formulae for ψ and ω, he chose a boundary integration path that closely skirted the singular point, thus introducing a source of error.

Because the shape of the free surface must be found from some iterative process and is typically non-trivial, SA BIMs are unusable for creeping planar extrusion flows with $0 < Ca < \infty$. It is not possible for these flows to employ SS BIMs either: the angle of separation being unknown a priori means that the exponents in the leading order asymptotic form g (see (1)) must be numerically calculated and it is not known how to do this for more than the first two exponents, meaning that the flow variables will not have the required regularity throughout the flow domain. In this article, we use a novel SI BIM, a modification of the BIM of Hansen and Kelmanson [10], [11], to solve the problem of the planar extrusion of a Newtonian fluid at vanishing Reynolds number Re and, in particular, to determine the shape of the free surface in the immediate neighbourhood of the separation point C for a range of capillary numbers.

The relevance of the results of our calculations to steady extrusion flows at non-zero Reynolds numbers, may be understood from the following argument of Moffatt [8]: Let us suppose that in the neighbourhood of C the stream function for the two-dimensional Navier–Stokes equations may be written in terms of local polar coordinates (r, θ) in the form

$$\psi \sim A r^{1+\lambda} f(\theta), \tag{2}$$

for some constant A, exponent λ and function f. Then, denoting the fluid velocity by $\boldsymbol{v}$ we see that the inertial term $\boldsymbol{v} \cdot \nabla \boldsymbol{v} = \mathcal{O}(A^2 r^{2\lambda-1})$ is negligible compared to the viscous term $\nu \nabla^2 \boldsymbol{v} = \mathcal{O}(\nu A r^{\lambda-2})$ if

$$\frac{A r^{\lambda+1}}{\nu} \ll 1. \tag{3}$$

The inequality (3) holds (provided $Re(\lambda) > 0$) for r sufficiently small. That is, in the neighbourhood of the singularity we expect the behaviour to be Stokesian. Neither is the local analysis near the separation point C in the present study limited to the strictly planar case. At zero Reynolds numbers, and in the absence of gravity, the (three-dimensional) flow of a Newtonian fluid extruded from a circular die will be axisymmetric and sufficiently far from the die lip there are obvious differences between the three-dimensional and planar extrudate. Such differences would include the values of the extrudate swell ratios computed under different flow conditions [19]. Ramalingam [20] points out, however, that the analysis of the singular behaviour of the flow variables near the orifice of an axisymmetric extrusion flow may be expected to show it to be locally planar. See too an analysis of Burda [21] of the asymptotics near corners in axisymmetric Stokes flow. The novel SI BIM presented in the present article could therefore, we believe, be adapted to the axisymmetric case.

1.2 Outline

The outline of this article is as follows: After briefly presenting the governing equations for Stokes flow in Section 2.1 we proceed to give a concise mathematical description of the problem to be solved, together with the boundary conditions that apply in Section 2.2. In Section 2.3 we describe how the mathematical equation of the free surface is divided into two parts and how the two surfaces are matched at some small distance from the separation point. The form of the surface near the separation point C has slope $\tan(\alpha)$ (for an angle α which must be calculated) and allows for the possibility of infinite curvature there. Further from C we use a linear combination of linearly independent functions proposed by Kelmanson [2] whose parameters must be adjusted in an attempt to satisfy the normal force balance condition on the free surface.

In Section 2.4, and inspired by Michael [13] and Lugt and Schwiderski [22], we derive the singular form of the stream function ψ, the vorticity ω and the pressure p in the neighbourhood of the separation point. It may be concluded that the integrals representing ψ and ω in the classical BIM diverge if the integration path passes through the separation point because of the non-integrability of $\partial \omega / \partial n$ there.

In Section 3 we outline two different BIMs for the solution of the boundary value problem of interest and explain how they differ. In Section 3.1 the BIM of Hansen and Kelmanson [10], [11], already mentioned in the Introduction, is treated to a more thorough exposition. Then, in Section 3.2 we show that it is not necessary to use integration by parts on the entire boundary when $\partial \omega / \partial n$ is replaced by $\partial p / \partial s$ in the integral representations of ψ and ω, but that it may be confined to the intersection of the boundary with a small neighbourhood of the separation point. Since the integration path used in the present BIM passes directly through the separation point this is an important difference between the new method and that of Owens [14] who was obligated to skirt the singularity. In supplying details of the discretization of the new BIM in Section 3.2.1 it becomes clear that there may be fewer jump terms in the pressure to contend with than in Hansen and Kelmanson's BIM and that the boundary grid for the computation of p and the other flow variables becomes a staggered one. We enumerate some of the other advantages of the present method with respect to that of Hansen and Kelmanson in Section 3.2.2.

The parameters appearing in the linear combination of functions proposed by Kelmanson [2] for the representation of the free surface sufficiently far from the separation point are calculated using the Levenberg–Marquardt [23], [24] method which solves the non-linear least-squares problem arising from the minimisation of the residual of the normal stress balance at certain selected points on the free surface. We explain this is more detail in Section 3.3.

Some numerical results, obtained on three different meshes, are presented in Section 4. It is shown in Section 4.1 that for any given surface tension, the (positive) slope of the free surface attains a maximum near the separation point and that the second derivative grows dramatically as the separation point is approached along the free surface, the curvature at the separation point itself being infinite. In Section 4.2 we present results for the extrudate swell ratio over a range of capillary numbers and these are shown to be in excellent agreement with others in the numerical literature [14], [17], [19], [25]. However, recognising that the extrudate swell ratio is a somewhat crude measure of the correctness of the solution, in Section 4.3 we also post our calculated values for the separation angles α and leading order indices λ_1 appearing in the asymptotic forms of the flow variables in the immediate neighbourhood of the separation point. The dearth and spread of data for these two parameters in the scientific literature, be it theoretical [16], numerical [17] or experimental [26]–[29] make comparisons difficult although agreement between our results with what has been previously published would seem to be satisfactory.

2 PLANAR EXTRUSION OF A NEWTONIAN FLUID AT VANISHING REYNOLDS NUMBER

2.1 Governing equations for steady planar Stokes flow

The governing non-dimensional equations for steady Stokes flow in a domain Ω are

$$\nabla \cdot \boldsymbol{v} = 0, \qquad \text{in } \Omega, \tag{4}$$

$$-\nabla p + \Delta \boldsymbol{v} = \boldsymbol{0}, \qquad \text{in } \Omega, \tag{5}$$

where $\boldsymbol{v}$ is the fluid velocity and p is the pressure. In two dimensions, (4) allows us to introduce a stream function ψ such that $\boldsymbol{v} = (u(x,y), v(x,y))$ may be written

$$\boldsymbol{v} = \nabla \times (0,0,\psi) \Rightarrow u = \frac{\partial \psi}{\partial y} \text{ and } v = -\frac{\partial \psi}{\partial x}, \tag{6}$$

where, in the usual notation, x and y are dimensionless Cartesian coordinates. By calculating the curl of (5) we get the biharmonic equation for ψ:

$$\Delta^2 \psi = 0. \tag{7}$$

We now introduce $\omega := \Delta \psi$, and see that (7) may be re-expressed as a system of coupled equations

$$\begin{cases} \Delta \omega = 0, & \text{in } \Omega, \\ \Delta \psi = \omega, & \text{in } \Omega. \end{cases} \tag{8}$$

2.2 Geometry, problem description and boundary conditions

2.2.1 Problem geometry and assumptions
The upper portion of the two-dimensional geometry of the problem to be solved is shown in Fig. 1. Dimensionless Cartesian coordinates (x, y) are used and x is chosen, in a natural way,

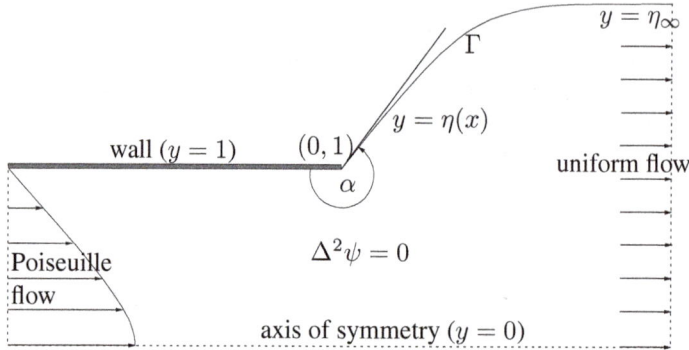

Figure 1: Problem geometry of planar extrusion.

as the stream-wise coordinate. Far upstream a Newtonian fluid is driven, at negligibly small Reynolds number, by a constant dimensionless pressure gradient $\partial p / \partial x$ along the channel formed between two parallel walls $y = \pm 1$ giving rise to steady planar Poiseuille flow. At $x = 0$ the fluid is extruded into air and the points $(0, \pm 1)$ are separation points where the fluid undergoes a transition from being confined by the walls to forming a jet. The tangent to the free surface Γ at the point of separation $(0, 1)$ forms an angle α with the solid wall $y = 1$, as shown in Fig. 1. Sufficiently far downstream of the line of extrusion into air the flow is considered to be steady and uniform and the free surface is parallel to the $x-$axis, having equation $y = \eta_{\infty}$ for some constant η_{∞}, called the extrudate swell ratio.

2.2.2 Boundary conditions for ψ and ω

The coupled eqn (8) are solved subject to boundary conditions derived by further assuming no-slip and no-penetration conditions along the solid wall $\{x \leq 0,\, y = 1\}$, unit volume flow rate in the upper half channel far upstream, and zero shear and normal stresses along the free surface Γ and axis of symmetry $y = 0$.

At entry $(x \to -\infty)$ we have

$$\psi = \frac{3}{2} y \left(1 - \frac{y^2}{3} \right), \quad \omega = -3y, \tag{9}$$

whereas at exit $(x \to \infty)$ uniform flow and incompressibility lead to

$$\psi = \frac{y}{\eta_{\infty}}, \quad \omega = 0. \tag{10}$$

No-slip and no-penetration conditions on the upper solid wall $\{x \leq 0,\, y = 1\}$ are ensured by prescribing

$$\psi = 1, \quad \frac{\partial \psi}{\partial y} = 0, \tag{11}$$

and along the axis of symmetry $(y = 0)$ the zero shear-rate and no-penetration conditions give rise to

$$\psi = 0, \quad \frac{\partial^2 \psi}{\partial y^2} = 0. \tag{12}$$

Along the free surface Γ, continuous with the solid wall, we must now have

$$\psi = 1. \tag{13}$$

Let $\partial^2/\partial t^2$ denote the second-order tangential derivative. Then, the zero shear-stress condition on the free surface

$$\frac{\partial^2 \psi}{\partial n^2} - \frac{\partial^2 \psi}{\partial t^2} = 0, \tag{14}$$

can be used to show (see, for example, Batchelor [30] or Longuet-Higgins [31]) that on Γ

$$\omega = -2\kappa \frac{\partial \psi}{\partial n}, \tag{15}$$

where κ is the signed curvature, defined by

$$\kappa = -\nabla \cdot \hat{\boldsymbol{n}}, \tag{16}$$

and $\hat{\boldsymbol{n}}$ is the unit outward pointing normal vector on Γ. Finally, a normal force balance on the free surface gives

$$[p] + 2\frac{\partial^2 \psi}{\partial n \partial s} + \gamma \kappa = 0, \tag{17}$$

where the excess pressure $[p] := p - p_a$ is the difference between the fluid pressure p and atmospheric pressure p_a, γ is the surface tension and $\partial/\partial s$ ($= \partial/\partial t$) denotes a derivative with respect to arc length along Γ. From this point on, and to ease the notation, we will simply denote the excess pressure by p.

2.3 Mathematical description of the free surface

The mathematical equation of the free surface Γ is divided into two parts and the two surfaces are matched at some small distance $r_c > 0$ from the separation point C. The free surface at the separation point has slope $\tan(\alpha)$ and we allow for the possibility of infinite curvature there. Further from C we use a linear combination of linearly independent functions proposed by Kelmanson [2] whose parameters must be adjusted in an attempt to satisfy the normal force balance condition (17).

2.3.1 At a distance $r \geq r_c$ from the separation point C

Sufficiently far from the separation point the equation of the free surface Γ is written $y = \eta(x)$, where

$$\eta(x) = \sum_{i=1}^{n} \beta_i \, \eta_i(x), \tag{18}$$

and where the n linearly independent functions $\{\eta_i\}_{i=1}^{n}$ are defined by

$$\eta_i(x) = 1 + \alpha_i \tanh\left(x\left(\varepsilon_{\infty,i} - (\varepsilon_{0,i} - \varepsilon_{\infty,i})\exp(-\gamma_i x)\right)\right), \qquad i = 1, 2, \ldots, n. \tag{19}$$

The parameters $\{\alpha_i, \varepsilon_{\infty,i}, \varepsilon_{0,i}, \gamma_i\}$ appearing in (19), together with the $\{\beta_i\}_{i=1}^{n}$ of (18) are to be determined. Although the form (18) is not used at C we nevertheless impose $\eta(0) = 1$. This constraint means that

$$\sum_{i=1}^{n} \beta_i = 1, \tag{20}$$

which reduces the number of parameters in (18) that need to be calculated to $5n - 1$.

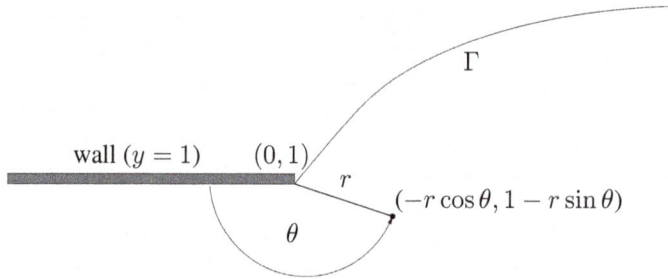

Figure 2: Polar angle θ measured in the anti-clockwise direction from the channel wall $y = 1$.

2.3.2 At a distance $r \leq r_c$ from the separation point C

In the sector centred at C and of radius r_c we express the equation of the free surface in polar coordinates (again, see Fig. 2) as the truncated series

$$y = 1 + h(r) = 1 - r \sin\alpha + h_1\, r^{\lambda_1 + 1} + h_2\, r^{2\lambda_1 + 1} - \frac{c_p\, r^2}{2\gamma} \cos\alpha, \qquad (21)$$

where the parameters $\{\alpha, \lambda_1, h_1, h_2, c_p\}$ are found so that to leading orders the curve Γ and its curvature might be continuous at $r = r_c$, so that non-trivial singular solutions (see Section 2.4) may be found for all variables in the sector of radius r_c centred at C and so that to leading orders the normal stress balance (17) may be satisfied on the surface (21).

2.4 Singular representations

We now seek a solution to the biharmonic equation (7) in the near neighbourhood of the separation point. One can show (see, for example, Michael [13] or Lugt and Schwiderski [22]) that in a sector centred at $(0, 1)$ and provided that $\lambda_n \neq 1$ and $Re(\lambda_n) > 0$ $(n = 1, 2, \ldots)$, ψ may be written in the form

$$\psi = 1 + \sum_{n=1}^{\infty} r^{\lambda_n + 1} f_n(\theta), \qquad (22)$$

where r and θ are polar coordinates centred at $(0, 1)$ (see Fig. 2), the eigenvalues λ_n are to be determined and

$$f_n(\theta) = A_n \cos\left((\lambda_n + 1)\theta\right) + B_n \sin\left((\lambda_n + 1)\theta\right)$$
$$+ C_n \cos\left((\lambda_n - 1)\theta\right) + D_n \sin\left((\lambda_n - 1)\theta\right), \qquad (23)$$

where $\{A_n, B_n, C_n, D_n\}$ $(n = 1, 2, 3, \ldots)$ are coefficients to be calculated. The radial and transverse components of the velocity v are given by

$$v_r = \frac{1}{r}\frac{\partial\psi}{\partial\theta} = \sum_{n=1}^{\infty} r^{\lambda_n} f_n'(\theta) \text{ and } v_\theta = -\frac{\partial\psi}{\partial r} = -\sum_{n=1}^{\infty} (\lambda_n + 1) r^{\lambda_n} f_n(\theta), \qquad (24)$$

so that, as observed for example by Sturges [32], the weak regularity condition $Re(\lambda_n) > 0$ means that the velocity vanishes as $r \to 0$. Following Michael [13], we also observe that

$Re(\lambda_n) > 0$ is a physical requirement in order that the integrated stresses remain finite as $r \to 0$.

The boundary conditions (11), (13) and (14) may be shown to give rise to a homogeneous linear system of equations for A_1, B_1, C_1 and D_1 which, provided that it is singular, allows the determination of three of these coefficients in terms of the fourth. In the present paper we choose to find A_1, B_1 and D_1 in terms of C_1. The boundary conditions of Section 2.2.2 also allow A_2, B_2, C_2 and D_2 to be written in terms of C_1. In agreement with the r exponent in the correction $Ca\psi_1$ to the leading order term in the perturbation expansion (2.13a) of the stream function in an article of Anderson and Davis [15], it may be shown [14] that $\lambda_2 = 2\lambda_1$ is necessary in order that the separation angle α be different from π. Thus, for $i = 1, 2$ we may write

$$f_i(\theta) = C_1 F_i(\theta), \qquad i = 1, 2,$$

with the function F_i having as parameters only λ_1 and α. Truncating (22) after two terms we are therefore left with

$$\psi(r, \theta) \sim 1 + C_1 \Psi(r, \theta), \tag{25}$$

where we define

$$\Psi(r, \theta) := r^{\lambda_1 + 1} F_1(\theta) + r^{2\lambda_1 + 1} F_2(\theta). \tag{26}$$

A singular expression for p in the neighbourhood of the separation point may be obtained using (5) and the relationship (6) between v and ψ:

$$p \sim C_1 \Pi(r, \theta) - c_p, \tag{27}$$

where

$$\Pi(r, \theta) := \frac{r^{\lambda_1 - 1}}{\lambda_1 - 1}((\lambda_1 + 1)^2 F_1'(\theta) + F_1^{(3)}(\theta)) + \frac{r^{2\lambda_1 - 1}}{2\lambda_1 - 1}((2\lambda_1 + 1)^2 F_2'(\theta) + F_2^{(3)}(\theta)), \tag{28}$$

and c_p is a constant in the definition of $y = 1 + h(r)$ (see eqn (21)).

Finally, using $\omega = \Delta\psi$ (see (8)) we get an expression for ω in the neighbourhood of the separation point:

$$\omega(r, \theta) \sim C_1 \Omega(r, \theta), \tag{29}$$

where

$$\Omega(r, \theta) := (1 + \lambda_1)^2 \, r^{\lambda_1 - 1} F_1(\theta) + (1 + 2\lambda_1)^2 \, r^{2\lambda_1 - 1} F_2(\theta). \tag{30}$$

The coefficient C_1 in (25), (27) and (29) is to be determined. λ_1 and α appearing in the functions F_1 and F_2 and c_p a constant in (27) are found from a nonlinear least-squares calculation of the free surface, as will be described in Section 3.3. One can show (see, for example, Table 1 of Sturges [32]) that if $\alpha \in [\pi, 3\pi/2]$ the value of λ_1 will be real and remain in the interval $[1/3, 1/2]$. Therefore, from (26), (28) and (30) whereas $\psi, \partial\psi/\partial n, p$ and ω are all Riemann integrable at $r = 0$, $\partial\omega/\partial n = \mathcal{O}(r^{\lambda_1 - 2})$ is not.

3 BOUNDARY INTEGRAL FORMULATIONS FOR THE SOLUTION OF (7)

Let Ω be the truncated domain whose boundary $\partial\Omega$ is $AB \cup BC' \cup C'C'_\rho \cup \mathscr{C}_\rho \cup C''_\rho C'' \cup C''D \cup DE \cup EA$, as shown in Figs 3 and 4. The inflow boundary AB is set at a finite distance $-x_{-\infty}$ upstream and the outflow boundary DE a finite distance x_∞ downstream of the die exit. C is the separation point $(0, 1)$ and the points C' and C'' are both at a distance r_c from C on the wall BC and the free surface CD, respectively. Thus, C' is the point $(-r_c, 1)$

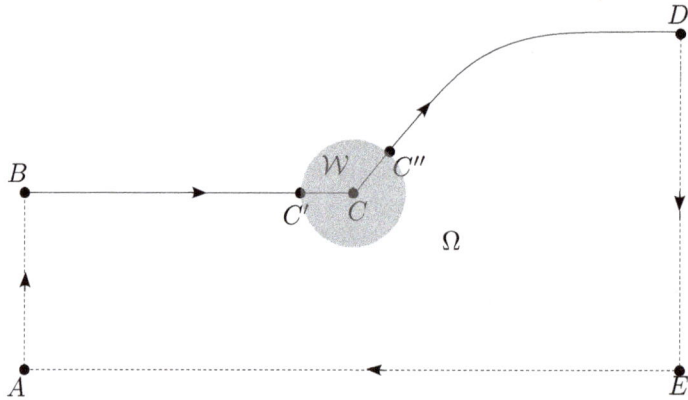

Figure 3: The truncated flow domain $ABCDE$ and the singular region $\mathcal{W} \cap \overline{\Omega}$..

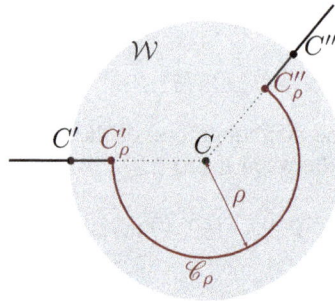

Figure 4: Figure showing the integration path $C'C'_\rho \cup \mathscr{C}_\rho \cup C''_\rho C''$ contained in the singular region $\mathcal{W} \cap \overline{\Omega}$.

and C'' the point $(x_0, \eta(x_0))$ where x_0 is the positive root of

$$x_0^2 + (\eta(x_0) - 1)^2 = r_c^2. \tag{31}$$

In Figs 3 and 4, the points C'_ρ and C'' are points on the wall BC and the free surface CD, respectively, a distance $0 < \rho \le r_c$ from C. $\mathscr{C}_\rho$ is the arc of radius ρ centred at C, having initial point C'_ρ and end point C''_ρ, as shown in Fig. 4. $\mathcal{W}$ denotes the disk of radius r_c centred at C and in the sector $\mathcal{W} \cap \overline{\Omega}$ the singular forms (25), (27) and (29) or their derivatives will be used to represent all flow variables appearing in (32), (33) and (39).

Suppose that $P(x', y')$ is an arbitrary point in $\mathbb{R}^2$. Then, using Green's identities it can be shown that $\psi(x', y')$ and $\omega(x', y')$, the solutions to (8), may be expressed in integral form as

$$\xi(x', y')\omega(x', y') = \int_{\partial\Omega} \left(\frac{\partial\omega}{\partial n} G_1 - \omega \frac{\partial G_1}{\partial n} \right) ds, \tag{32}$$

$$\xi(x', y')\psi(x', y') = \int_{\partial\Omega} \left(\frac{\partial\omega}{\partial n} G_2 - \omega \frac{\partial G_2}{\partial n} + \frac{\partial\psi}{\partial n} G_1 - \psi \frac{\partial G_1}{\partial n} \right) ds, \tag{33}$$

where the normal derivative $\partial/\partial n := \hat{\boldsymbol{n}} \cdot \nabla$ and $\hat{\boldsymbol{n}}$ is an outward pointing unit normal vector to $\partial\Omega$. The factor $\xi(x', y')$ in (32) and (33) is defined by

$$\xi(P) = \begin{cases} \dfrac{\varphi}{2\pi} & \text{if } P \in \partial\Omega, \\ 1 & \text{if } P \in \Omega, \\ 0 & \text{otherwise,} \end{cases} \tag{34}$$

where φ is the angle between the left and right tangents at the boundary point P when $P \in \partial\Omega$. The functions $G_1 = G_1(x, y, x', y')$ and $G_2 = G_2(x, y, x', y')$ appearing in (32) and (33) are fundamental solutions of, respectively, the two-dimensional Laplace's equation and biharmonic equation, chosen to be

$$G_1 = -\frac{1}{4\pi} \log((x - x')^2 + (y - y')^2), \tag{35}$$

$$G_2 = -\frac{1}{16\pi}(((x - x')^2 + (y - y')^2)(\log((x - x')^2 + (y - y')^2) - 2)). \tag{36}$$

Along a plane piecewise smooth curve $\mathcal{C}$ it may be easily shown from (5) and (6) that

$$\frac{\partial p}{\partial s} = \frac{\partial \omega}{\partial n}, \tag{37}$$

almost everywhere, a result that will be seen to be of central importance in the sections that follow.

3.1 The formulation of Hansen and Kelmanson [10], [11]

The approach of Hansen and Kelmanson [10], [11] uses integration by parts to write, along a smooth finite curve $\mathcal{C}$ having initial point a and end point b

$$\int_{\mathcal{C}} \frac{\partial \omega}{\partial n} G \, ds = \int_{\mathcal{C}} \frac{\partial p}{\partial s} G \, ds = pG(b) - pG(a) - \int_{\mathcal{C}} p \frac{\partial G}{\partial s} \, ds, \tag{38}$$

where G is some suitable differentiable function and it has been assumed that pG is continuous along $\mathcal{C}$. Setting $G = G_1$ (35) or G_2 (36), the generalization of (38) to the simple closed piecewise smooth boundary curve $\mathcal{C} = \partial\Omega$, where pG has a finite number of jumps, is straightforward and, in the present case, leads to

$$\int_{\partial\Omega} \frac{\partial p}{\partial s} G_j \, ds = \int_B^{C'_\rho} \frac{\partial p}{\partial s} G_j \, ds + \int_{\mathscr{C}_\rho} \frac{\partial p}{\partial s} G_j \, ds + \int_{C''_\rho}^D \frac{\partial p}{\partial s} G_j \, ds + \int_E^A \frac{\partial p}{\partial s} G_j \, ds$$

$$= [pG_j]_B^{C'-} + [pG_j]_{C'+}^{C'_\rho} + [pG_j]_{C'_\rho}^{C''_\rho} + [pG_j]_{C''_\rho}^{C''-} + [pG_j]_{C''+}^D + [pG_j]_E^A$$

$$- \int_B^{C'_\rho} p \frac{\partial G_j}{\partial s} \, ds - \int_{\mathscr{C}_\rho} p \frac{\partial G_j}{\partial s} \, ds - \int_{C''_\rho}^D p \frac{\partial G_j}{\partial s} \, ds - \int_E^A p \frac{\partial G_j}{\partial s} \, ds$$

$$= -[pG_j]_A^B - [pG_j]_{C'-}^{C'+} - [pG_j]_{C''-}^{C''+} - \int_B^{C'_\rho} p \frac{\partial G_j}{\partial s} \, ds - \int_{\mathscr{C}_\rho} p \frac{\partial G_j}{\partial s} \, ds$$

$$- \int_{C''_\rho}^D p \frac{\partial G_j}{\partial s} \, ds - \int_E^A p \frac{\partial G_j}{\partial s} \, ds, \quad j = 1, 2. \tag{39}$$

We assume here and henceforth that (x', y') is not chosen to coincide with the separation point. Thus, by the integrability of $p\partial G_j/\partial s$ along $\mathscr{C}_\rho$ for all $\rho > 0$ we have, in the limit $\rho \to 0$,

$$\int_{\partial\Omega} \frac{\partial p}{\partial s} G_j \, ds = -[pG_j]_A^B - [pG_j]_{C'-}^{C'+} - [pG_j]_{C''-}^{C''+} - \int_B^D p\frac{\partial G_j}{\partial s} \, ds - \int_E^A p\frac{\partial G_j}{\partial s} \, ds,$$
(40)

where it is understood (since $C'_\rho, C''_\rho \to C$ when $\rho \to 0$) that

$$\int_B^D p\frac{\partial G_j}{\partial s} \, ds = \int_B^{C'} p\frac{\partial G_j}{\partial s} \, ds + \int_{C'}^C p\frac{\partial G_j}{\partial s} \, ds + \int_C^{C''} p\frac{\partial G_j}{\partial s} \, ds + \int_{C''}^D p\frac{\partial G_j}{\partial s} \, ds.$$
(41)

Thus, (40) becomes

$$\int_{\partial\Omega} \frac{\partial p}{\partial s} G_j \, ds = -[pG_j]_A^B - [pG_j]_{C'-}^{C'+} - [pG_j]_{C''-}^{C''+} - \int_B^{C'} p\frac{\partial G_j}{\partial s} \, ds - \int_{C'}^C p\frac{\partial G_j}{\partial s} \, ds$$
$$- \int_C^{C''} p\frac{\partial G_j}{\partial s} \, ds - \int_{C''}^D p\frac{\partial G_j}{\partial s} \, ds - \int_E^A p\frac{\partial G_j}{\partial s} \, ds.$$
(42)

In [10], [11], Hansen and Kelmanson's boundary integral discretizations of eqns (32) and (33) used midpoint collocation for the pressure and vorticity in each of the subintervals outside a singular region.

3.2 A new formulation

Rather than perform an integration by parts all along $\partial\Omega$, as in (38), we propose to do so only along that part of $\partial\Omega$ that is within $\mathcal{W} \cap \overline{\Omega}$ (where all variables have a singular representation) and then let $\rho \to 0$. (Of course, another possibility would be to retain $\partial\omega/\partial n G_j$ ($j = 1, 2$) outside $\mathcal{W} \cap \overline{\Omega}$ but then if the pressure is subsequently needed on the boundary it will have to be determined in a post-processing step by integrating (37) along $\partial\Omega$.) First of all,

$$\int_{\partial\Omega} \frac{\partial p}{\partial s} G_j \, ds = \int_B^{C'} \frac{\partial p}{\partial s} G_j \, ds + \left(\int_{C'}^{C'_\rho} \frac{\partial p}{\partial s} G_j \, ds + \int_{\mathscr{C}_\rho} \frac{\partial p}{\partial s} G_j \, ds + \int_{C''_\rho}^{C''} \frac{\partial p}{\partial s} G_j \, ds \right)$$
$$+ \int_{C''}^D \frac{\partial p}{\partial s} G_j \, ds + \int_E^A \frac{\partial p}{\partial s} G_j \, ds.$$
(43)

We now use integration by parts on the terms inside the parentheses in (43) and then let $\rho \to 0$ to get

$$\lim_{\rho\to 0} \left([pG_j]_{C'}^{C'_\rho} + [pG_j]_{C'_\rho}^{C''_\rho} + [pG_j]_{C''_\rho}^{C''} - \int_{C'}^{C'_\rho} p\frac{\partial G_j}{\partial s} \, ds - \int_{\mathscr{C}_\rho} p\frac{\partial G_j}{\partial s} \, ds - \int_{C''_\rho}^{C''} p\frac{\partial G_j}{\partial s} \, ds \right)$$
$$= \lim_{\rho\to 0} \left(pG_j(C''-) - pG_j(C'+) - \int_{C'}^{C'_\rho} p\frac{\partial G_j}{\partial s} \, ds - \int_{\mathscr{C}_\rho} p\frac{\partial G_j}{\partial s} \, ds - \int_{C''_\rho}^{C''} p\frac{\partial G_j}{\partial s} \, ds \right)$$
$$= [pG_j]_{C'+}^{C''-} - \int_{C'}^C p\frac{\partial G_j}{\partial s} \, ds - \int_C^{C''} p\frac{\partial G_j}{\partial s} \, ds,$$
(44)

where, again, $(x', y') \neq (0, 1)$ and the integrability of $p\partial G_j/\partial s$ along $\mathscr{C}_\rho$ for all $\rho > 0$ has allowed us to write

$$\lim_{\rho\to 0} \int_{\mathscr{C}_\rho} p\frac{\partial G_j}{\partial s} \, ds = 0.$$

In summary, (43) becomes

$$
\int_{\partial\Omega} \frac{\partial p}{\partial s} G_j \, ds = \left[p \, G_j \right]_{C'+}^{C''-} + \int_B^{C'} \frac{\partial p}{\partial s} G_j \, ds - \int_{C'}^{C} p \frac{\partial G_j}{\partial s} \, ds - \int_{C}^{C''} p \frac{\partial G_j}{\partial s} \, ds
$$

$$
+ \int_{C''}^{D} \frac{\partial p}{\partial s} G_j \, ds + \int_E^{A} \frac{\partial p}{\partial s} G_j \, ds. \quad (45)
$$

3.2.1 Discretization

Referring throughout this section to Fig. 5, we now discretize the integrals appearing on the right-hand sides of eqns (32) and (33) where, in each case, the terms involving the integrals of $\partial\omega/\partial n$ are transformed into integrals involving the pressure p, as described in eqns (43)–(45). First of all, using the boundary conditions (9) and (10) and the assumption that the flow at both entry and exit is fully developed (so that, for example, $\partial p/\partial s = 0$) we are able to evaluate the contributions along AB and DE to the integrals in (32) and (33) exactly. Then, exploiting the boundary conditions (11), (12), (13) and (15) reduces the flow variables to be determined on the boundary to just p and ω along BC and to p and $\partial\psi/\partial n$ along $CD \cup EA$ as shown in Fig. 5. Since $\partial\psi/\partial n = 0$ on the solid wall the third term in the contribution along BC to the integral on the right-hand side of (33) vanishes. Further simplification follows from the knowledge of ψ along the upper and lower boundaries. In the remainder of this section we detail the discretization of the right-hand side of (33). The treatment of the integral in (32) is analogous.

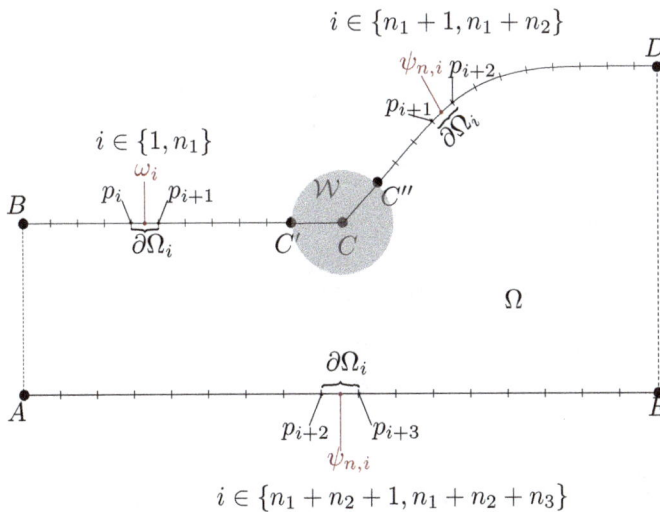

Figure 5: Summary of discrete variables along the boundary $\partial\Omega$, showing the division into $n_1 + n_2 + n_3$ sub-intervals $\partial\Omega_i$.

Along BC' The part BC' of the boundary is written as the union of n_1 subintervals $\partial\Omega_i$, $(i = 1, 2, \ldots, n_1)$ the length of the ith subinterval being denoted $\Delta x_i > 0$:

$$BC' = \bigcup_{i=1}^{n_1} \partial\Omega_i, \text{ with } |BC'| = \sum_{i=1}^{n_1} \Delta x_i. \tag{46}$$

The contribution along BC' to the integral on the right-hand side of (33) therefore becomes

$$\int_B^{C'} \left(\frac{\partial p}{\partial s} G_2 - \omega \frac{\partial G_2}{\partial n} - \frac{\partial G_1}{\partial n} \right) ds = \sum_{i=1}^{n_1} \int_{\partial\Omega_i} \left(\frac{\partial p}{\partial s} G_2 - \omega \frac{\partial G_2}{\partial n} - \frac{\partial G_1}{\partial n} \right) ds, \tag{47}$$

and on each subinterval $\partial\Omega_i$ we approximate the integral by the midpoint rule

$$\int_{\partial\Omega_i} \left(\frac{\partial p}{\partial s} G_2 - \omega \frac{\partial G_2}{\partial n} - \frac{\partial G_1}{\partial n} \right) ds \approx \left(\frac{\partial p}{\partial s} \right)_i \int_{\partial\Omega_i} G_2 \, ds - \omega_i \int_{\partial\Omega_i} \frac{\partial G_2}{\partial n} \, ds - \int_{\partial\Omega_i} \frac{\partial G_1}{\partial n} \, ds, \tag{48}$$

where $(\partial p/\partial s)_i$ and ω_i are approximations to the values of $\partial p/\partial s$ and ω at the midpoint of $\partial\Omega_i$. The partial derivative $(\partial p/\partial s)_i$ is approximated using a central difference approximation:

$$\left(\frac{\partial p}{\partial s} \right)_i \approx \frac{p_{i+1} - p_i}{\Delta x_i}, \tag{49}$$

where p_{i+1} and p_i are approximations to p at the end points of the subinterval $\partial\Omega_i$ (see Fig. 5). The unknowns to be determined from this contribution to the boundary integral on the right-hand side of (33) are therefore n_1 approximations to ω at the midpoints of the subintervals $\partial\Omega_i$ and $n_1 + 1$ approximations to p at the end points of these subintervals. Clearly, the final term on the right-hand side of (47), involving only $\partial G_1/\partial n$ passes to the right-hand side of the system to be solved for the unknowns.

Along $C''D$ In a similar fashion, we write $C''D$ as the union of n_2 subintervals $\{\partial\Omega_i\}_{i=n_1+1}^{n_1+n_2}$ and again use the midpoint rule and central difference approximations to get

$$\int_{C''}^D \left(\frac{\partial\omega}{\partial n} G_2 - \omega \frac{\partial G_2}{\partial n} + \frac{\partial\psi}{\partial n} G_1 - \psi \frac{\partial G_1}{\partial n} \right) ds$$

$$= \int_{C''}^D \left(\frac{\partial p}{\partial s} G_2 + \frac{\partial\psi}{\partial n} \left(2\kappa \frac{\partial G_2}{\partial n} + G_1 \right) - \frac{\partial G_1}{\partial n} \right) ds$$

$$\approx \sum_{i=n_1+1}^{n_1+n_2} \left(\frac{p_{i+2} - p_{i+1}}{\Delta s_i} \right) \int_{\partial\Omega_i} G_2 \, ds + \left(\frac{\partial\psi}{\partial n} \right)_i \int_{\partial\Omega_i} \left(2\kappa \frac{\partial G_2}{\partial n} + G_1 \right) ds - \int_{\partial\Omega_i} \frac{\partial G_1}{\partial n} \, ds, \tag{50}$$

where Δs_i denotes the arclength of the ith subinterval $\partial\Omega_i$. The unknowns to be determined from this contribution to the boundary integral are now, therefore, n_2 approximations to $\partial\psi/\partial n$ at the midpoints of the subintervals $\partial\Omega_i$ and $n_2 + 1$ approximations to p at the end points of these subintervals. As before, the final term on the right-hand side of (50), involving only $\partial G_1/\partial n$ passes to the right-hand side of the system to be solved for the unknowns.

Along EA The contribution along EA to the integral on the right-hand side of (33) is particularly simple due to the vanishing of both ω and ψ along the axis of symmetry. The decomposition of AE into n_3 subintervals $\{\partial\Omega_i\}_{i=n_1+n_2+1}^{n_1+n_2+n_3}$ and the use of the midpoint rule and central difference approximations leads to

$$\int_E^A \left(\frac{\partial p}{\partial s}G_2 + \frac{\partial \psi}{\partial n}G_1\right)ds \approx \sum_{i=n_1+n_2+1}^{n_1+n_2+n_3} -\left(\frac{p_{i+3}-p_{i+2}}{\Delta x_i}\right)\int_{\partial\Omega_i}G_2\,ds + \left(\frac{\partial\psi}{\partial n}\right)_i\int_{\partial\Omega_i}G_1\,ds,$$
(51)

where Δx_i denotes the length of $\partial\Omega_i$. The approximations to the pressure at the end points of the n_3 subintervals and the approximations to the normal derivative of the stream function at the midpoints of those subintervals represent the $(n_3 + 1 + n_3 = 2n_3 + 1)$ unknowns to be found from this contribution to the boundary integral.

Along $C'C \cup CC''$ Along this portion of the boundary $\partial\Omega$ we use the singular forms (27) for p, (29) for ω and the normal derivative of (25) for $\partial\psi/\partial n$ and integration by parts for the integral of $\partial p/\partial s\,G_2$, as explained in Section 3.2. Thus we have

$$[pG_2]_{C'+}^{C''-} + \int_{C'}^C \left(-p\frac{\partial G_2}{\partial s} - \omega\frac{\partial G_2}{\partial n} - \frac{\partial G_1}{\partial n}\right)ds$$
$$+ \int_C^{C''}\left(-p\frac{\partial G_2}{\partial s} - \omega\frac{\partial G_2}{\partial n} + \frac{\partial\psi}{\partial n}G_1 - \frac{\partial G_1}{\partial n}\right)ds$$
$$= [(C_1\Pi - c_p)G_2]_{C'+}^{C''-} + \int_{C'}^C\left(-(C_1\Pi - c_p)\frac{\partial G_2}{\partial s} - C_1\Omega\frac{\partial G_2}{\partial n} - \frac{\partial G_1}{\partial n}\right)ds$$
$$+ \int_C^{C''}\left(-(C_1\Pi - c_p)\frac{\partial G_2}{\partial s} - C_1\Omega\frac{\partial G_2}{\partial n} + C_1\Psi_nG_1 - \frac{\partial G_1}{\partial n}\right)ds$$
$$= C_1[\Pi G_2]_{C'+}^{C''-} + \int_{C'}^C\left(-C_1\Pi\frac{\partial G_2}{\partial s} - C_1\Omega\frac{\partial G_2}{\partial n} - \frac{\partial G_1}{\partial n}\right)ds$$
$$+ \int_C^{C''}\left(-C_1\Pi\frac{\partial G_2}{\partial s} - C_1\Omega\frac{\partial G_2}{\partial n} + C_1\Psi_nG_1 - \frac{\partial G_1}{\partial n}\right)ds. \quad (52)$$

The sole unknown from this contribution to the boundary integral is the singular coefficient C_1 and, as along BC' and $C''D$, the terms on the right-hand side of (52) involving $\partial G_1/\partial n$ pass to the right-hand side of the system to be solved for the unknowns. Note that for simplicity the approximate form $y = \eta(x)$ was used in the evaluation of the integrals along CC'' since the difference between the graph of this function and that of the singular form (21) is very small, even if the curvature of the two curves at the separation point is very different (one finite, one infinite).

Summary of flow variables to be determined For a given free surface Γ the above paragraphs should make it clear that a total of $2(n_1 + n_2 + n_3) + 4$ unknowns are to be calculated. To generate the same number of equations as unknowns we discretize, as explained above, eqns (32) and (33) with the choice of points (x', y') made equal to the Cartesian coordinates of the midpoints of the $n_1 + n_2 + n_3$ subintervals. This yields $2(n_1 + n_2 + n_3)$ equations. Assuming that the pressure is equal to that of the atmosphere at exit requires the prescription $p_{n_1+n_2+2} = p_{n_1+n_2+3} = 0$, supplying two more equations. The assumption that at inflow fully developed conditions apply and that therefore the pressure is a constant along AB means that we should set $p_1 = p_{n_1+n_2+n_3+3}$, giving one more equation. Finally, we choose one other point $(x', y') \neq (0, 1)$ different from all the others and satisfy the discrete version of one of eqns (32) and (33) to complete the required number of equations and create a square and non-singular coefficient matrix for the system to be solved. As is usual with boundary integral methods this matrix is full but sufficiently small (even for the finest meshes used) that an efficient direct solver may be used.

3.2.2 Differences with the formulation of Hansen and Kelmanson [10], [11]

Comparing eqns (42) and (45) in addition to the respective discretizations employed by Hansen and Kelmanson [10], [11] and proposed in this article in Section 3.2.1 reveals a number of important differences between the two approaches:

1. The present formulation leads to a somewhat simpler expression than (42) by avoiding jump terms at the points C' and C''.

2. The present formulation avoids what may be very delicate evaluations of improper integrals of $p\partial G_1(x, y, x', y')/\partial s$ along paths passing through the point (x', y'). Instead, we may have improper integrals of $\partial p/\partial s G_1$, but all Cauchy principal values exist and are simple to evaluate. More specifically, if a subinterval $\partial\Omega_i$ consists of an arc segment of length Δs_i then, as explained in Section 3.2.1, we use central differences to write

$$\int_{\partial\Omega_i} \frac{\partial p}{\partial s} G_1(x, y, x', y')\, ds \approx \left(\frac{p_{i+1} - p_i}{\Delta s_i}\right) \int_{\partial\Omega_i} G_1(x, y, x', y')\, ds. \qquad (53)$$

Let us denote the coordinates of the end points of $\partial\Omega_i$ by (x_i, y_i) and (x_{i+1}, y_{i+1}). Then, if $(x', y') \in \partial\Omega_i$ the Cauchy principal value of the integral appearing on the right-hand side of (53) is evaluated by expressing it as the sum of two convergent integrals: one over the subset of $\partial\Omega_i$ having end points (x_i, y_i) and (x', y') and the other over the subset of $\partial\Omega_i$ having end points (x', y') and (x_{i+1}, y_{i+1}). These are evaluated using an adaptive quadrature method: see Shampine [33].

3. When discretized (see Section 3.2.1) the new formulation will lead to a staggered boundary grid for the computation of p and ω along BC and of p and $\partial\psi/\partial n$ along $CD \cup EA$, the discrete pressure values being calculated at the end points of the subintervals $\{\partial\Omega_i\}_{i=1}^{n_1+n_2+n_3}$ and the remaining variables at the midpoints of these subintervals. This avoids possible problems with non-physical oscillatory pressure values or even pressure indeterminacy that may arise when all flow variables are collocated at the subinterval midpoints.

3.3 Determination of the free surface Γ

Thus far in the solution process, along $y = \eta(x)$ only the kinematic and shear-free conditions (13)–(15) have been satisfied and the free surface has been considered known. Of course, the correct free surface, along which the normal stress balance condition (17) is respected, must also be found. To find it we will use an iterative method and introduce the residual of the normal stress at $N + 1$ points along $C''D$, where we choose $N + 1 = 5n - 1$ in order to determine the parameters appearing in the linear combination (18) of the n basis functions η_i (see Section 2.3.1). We denote by R_i the residual of the normal stress balance at a point $(x_i, \eta(x_i))$ of the free surface:

$$R_i = p(x_i, \eta(x_i)) + 2\frac{\partial^2\psi}{\partial n\partial s}(x_i, \eta(x_i)) + \gamma\kappa(x_i, \eta(x_i)), \quad i = 1, 2, \ldots, N. \qquad (54)$$

R_i is approximated numerically using simple central differences in the calculation of the approximative derivative with respect to arc length of $\partial\psi/\partial n$ and spline interpolation of this and the calculated approximation to the pressure.

Next we use the Levenberg–Marquardt [23], [24] method, implemented in `lsqnonlin` of MATLAB, to solve the non-linear least-squares problem

$$\min_{\boldsymbol{c}} \sum_{i=1}^{N+1} R_i^2(\boldsymbol{c}), \tag{55}$$

where $\boldsymbol{c}$ is the vector containing the $5n-1$ parameters appearing in the linear combination (18) of the n basis functions η_i. Once the minimum has been found we recalculate C_1 and all the flow variables as detailed in Section 3.2.1 and continue until $\boldsymbol{c}^{(k+1)}$, the $(k+1)$th iterative value of $\boldsymbol{c}$, satisfies some convergence criterion, such as

$$\|\boldsymbol{c}^{(k+1)} - \boldsymbol{c}^{(k)}\|_2 < s\,\|1 + \boldsymbol{c}^{(k)}\|_2, \tag{56}$$

for some suitably small choice of the relative step tolerance s. Numerical continuation may be used to obtain solutions on increasingly refined meshes. That is to say, the converged vector $\boldsymbol{c}$ calculated for a certain choice of n_1, n_2 and n_3 may be used to calculate the initial free surface shape when finer meshes are employed.

4 NUMERICAL RESULTS

Calculations were performed for values of the dimensionless surface tension $\gamma = 0.001$, 0.01, 0.1 and 1 on meshes where the total number M of boundary elements $\partial\Omega_i$ varied from $M = 1434$ (2872 unknowns) to $M = 1854$ (3712 unknowns). The radius r_c of the singular sector $\mathcal{W} \cap \overline{\Omega}$ was set equal to 10^{-5}. With the choice of $n = 10$ basis functions η_i (see (19)) the number of points $N+1$ where the normal stress balance residual (54) was evaluated was equal to $5n - 1 = 49$. The Levenberg–Marquardt algorithm (see Section 3.3) was considered to have converged when the inequality (56) was satisfied with a relative step tolerance $s = 10^{-6}$. It was found that this choice of s always ensured that the final value of the objective function (55) was no more than $O(10^{-5})$.

4.1 Shape of the free surface

In Figs 6 to 8 we plot the converged free surface function $y = \eta(x)$, given by the linear combination (18), in addition to its first and second derivatives, over the full range of dimensionless surface tensions. It may be observed from Figs 7 and 8 that, for any given surface tension γ, the (positive) slope of the free surface attains a maximum at some point $0 < x_\gamma \ll 1$ and that the second derivative grows dramatically as $x \to 0$. These results were computed on the finest mesh ($M = 1854$) and in Section 3.3 we have explained how the parameters $\{\alpha_i, \varepsilon_{\infty,i}, \varepsilon_{0,i}, \gamma_i\}$ were calculated. Once they have been found it remains the case, however, that

$$\eta''(0) = -2\sum_{i=1}^{n} \beta_i \alpha_i (\epsilon_{0,i} - \epsilon_{\infty,i})\gamma_i,$$

is finite and that there is no reason to expect that

$$\eta'(0) = \sum_{i=1}^{n} \beta_i \alpha_i \epsilon_{0,i},$$

will equal $\tan(\alpha)$.

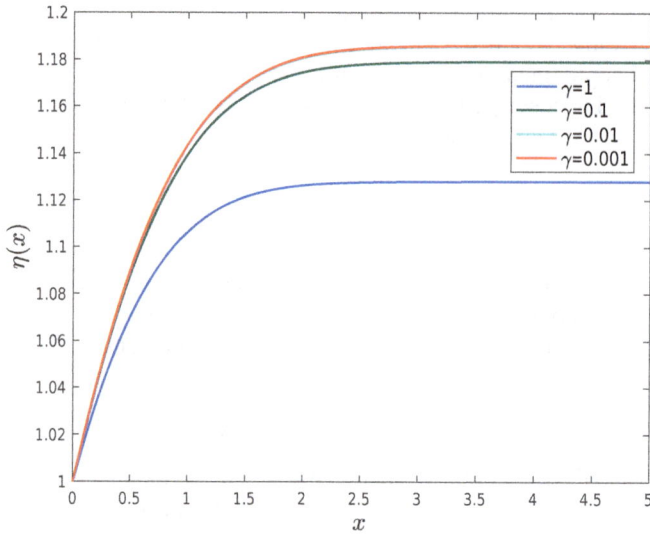

Figure 6: $n_1 = n_2 = n_3 = 618$. Free surface $y = \eta(x)$ computed at values 1, 0.1, 0.01 and 0.001 of the dimensionless surface tension γ.

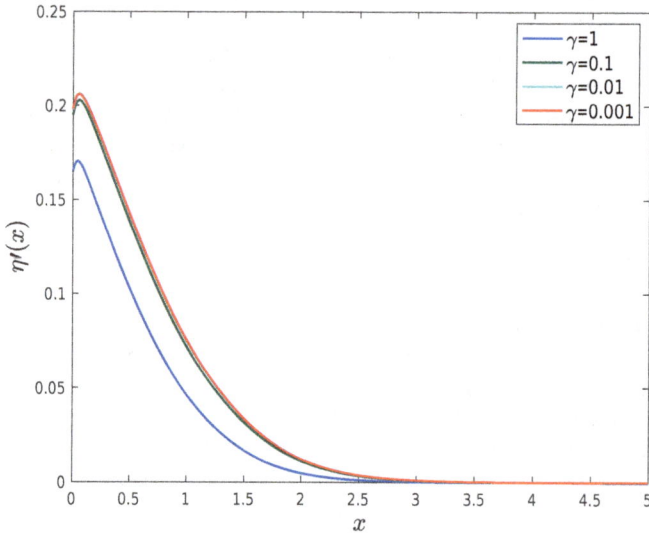

Figure 7: $n_1 = n_2 = n_3 = 618$. First derivative $y = \eta'(x)$ computed at values 1, 0.1, 0.01 and 0.001 of the dimensionless surface tension γ.

Given the deficiencies noted above of the representation (18) of the free surface in a neighbourhood of the separation point, we match the correct leading order asymptotic form (21), as explained in Section 2.3.2, to this far-field solution. In Fig. 9 we show the results of the matching at $x = x_0$ of (18) with (21), as computed on the $M = 1854$ mesh. Over

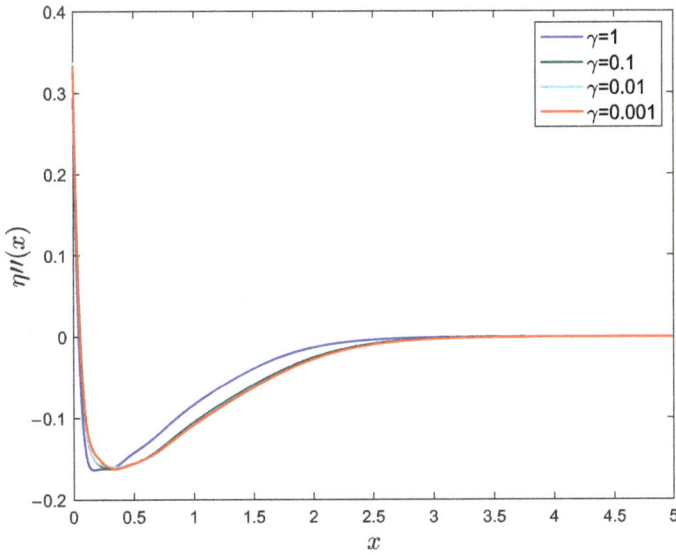

Figure 8: $n_1 = n_2 = n_3 = 618$. Second derivative $y = \eta''(x)$ computed at values 1, 0.1, 0.01 and 0.001 of the dimensionless surface tension γ.

$x \in [0, 2r_c]$ the graphs of (18) and (21) shown in Fig. 9 appear to the naked eye to be those of linear functions. In reality, $\kappa(0)$, the curvature at the origin, is infinite.

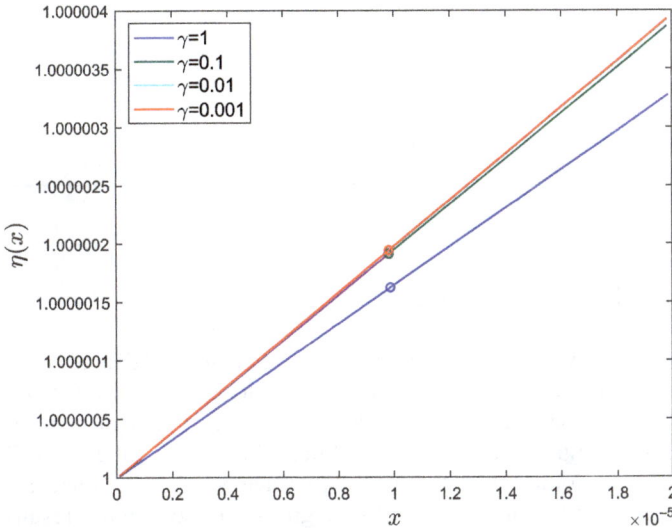

Figure 9: $n_1 = n_2 = n_3 = 618$. Matching of the inner solution $y = 1 + h(r)$ (21) to the outer solution $y = \eta(x)$ (18) at $x = x_0$ (see (31)) at values 1, 0.1, 0.01 and 0.001 of the dimensionless surface tension γ.

4.2 Extrudate swell ratio η_∞

Comparison of the calculated values of η_∞ on meshes with $M = 1434$, 1644 and 1854 is made in Table 1 with the results of the original boundary element method calculations with $M = 1854$ of Owens [14], with the result (at $\gamma = 1$) of the high-resolution finite element method of Salamon et al. [17], and with some of the results of the singular finite element calculations of Georgiou and Boudouvis [25] and of the finite element method of Mitsoulis et al. [19]. Since the result of Salamon et al. [17] was computed on a highly refined finite element mesh involving 171258 unknowns we believe that this may be considered to be a benchmark value for the case $\gamma = 1$. The value of the extrudate swell ratio at this value of γ computed by Georgiou and Boudouvis [25] is within 0.026% of the benchmark value and was obtained on a finite element mesh having 30866 unknowns. The results of Mitsoulis et al. [19] were obtained using meshes having either 22548 or 30866 degrees of freedom. Our calculation of η_∞ at $\gamma = 1$ using the $M = 1854$ boundary integral mesh is within 0.0629% of the value of Salamon et al. but was obtained using a mesh having only 3712 degrees of freedom. In this example, the cost advantages of using a boundary element method are clear. From the evidence presented in Table 1 it would seem that slightly higher values of η_∞ might be expected were we able to compute results on yet finer meshes, bringing them into even closer agreement with the results of Georgiou and Boudouvis [25], Mitsoulis et al. [19] and Salamon et al. [17]. Regrettably, however, calculations for $M > 1854$ are beyond what we can perform conveniently with the computing resources currently at our disposal.

Table 1:   Values of the extrudate swell ratio η_∞ for different values of the non-dimensional surface tension γ, computed using the boundary integral equation method and M boundary elements.

γ	Present method			Owens [14]	Georgiou and Boudouvis [25]	Mitsoulis et al. [19]	Salamon et al. [17]
	M=1434	M=1644	M=1854				
1	1.1279	1.1280	1.1281	1.1282	1.1291	1.129	1.12881
10^{-1}	1.1784	1.1793	1.1792	1.1792	1.1794	1.180	-
10^{-2}	1.1853	1.1853	1.1858	1.1857	-	1.186	-
10^{-3}	1.1859	1.1861	1.1861	1.1861	-	-	-
10^{-5}	-	-	-	-	-	1.186	-

4.3 Separation angles and leading order indices

In Table 2 we show the values of the parameters α (in radians and degrees) and λ_1 computed on the finest mesh and compare these with the results obtained by Owens [14] using a boundary integral method. We note in Table 2 a few very minor differences between the parameter values attributed to Owens [14] and those in his original article. This is because of small errors found in the calculation of the parameters A_2, B_2, C_2 and D_2 appearing in (23) since the article of Owens went to press. Agreement between our results and those of Owens [14] is excellent although the present results should be viewed as the more accurate of the two sets.

In the second and third columns of Table 2 we display the results for the separation angles in both radians and degrees when $\gamma = 1, 0.1, 0.01$ and 0.001, as computed on the finest mesh.

Following the ansatz by Schultz and Gervasio [16] for the form of the free surface $y = \eta(x)$ near the separation point:

$$\eta(x) - 1 = a + bx + cx^n, \tag{57}$$

for certain parameters a, b, c and n, Salamon et al. [17] estimated b to be equal to 0.176 when $\gamma = 1$ and $Re = 0$. Comparing (57) with (21) and noting that $r \approx x/\cos(\alpha - \pi)$ we can therefore estimate, when $\gamma = 1$, $\alpha \approx (180 + \arctan(0.176)) = 189.98°$, which is in satisfactory agreement with our own result of $189.37°$. Experimental results reported by Nickell et al. [26], Tanner [27] and Tanner et al. [28] were all for Reynolds numbers below 10^{-3} and those of Batchelor et al. [29] were for $Re \approx 10^{-8}$. The data from the experimental results reported by these authors showed separation angles to be between $189°$ and $194°$, so that our calculated values of α are well within the range of those measured experimentally.

There are very few results available in the literature for the index λ_1. Salamon et al. [17] estimated the value of the parameter n in the ansatz (57) to be 1.43 when $\gamma = 1$ and $Re = 0$, so that again, comparing (57) with (21) gives $\lambda_1 = n - 1 = 0.43$, not too far from the computed value shown in the first row and fourth column of Table 2.

Table 2: Values of the parameters α and λ_1 appearing in (21). The far-field solutions were computed on the finest mesh ($M = 1854$).

γ	Present method			Owens [14]†	
	α (radians)	α (degrees)	λ_1	α (radians)	λ_1
1	3.3051	189.3667	4.5317×10^{-1}	3.3049	4.5321×10^{-1}
10^{-1}	3.3342	191.0373	4.4587×10^{-1}	3.3337	4.4599×10^{-1}
10^{-2}	3.3368	191.1872	4.4522×10^{-1}	3.3363	4.4535×10^{-1}
10^{-3}	3.3371	191.2032	4.4516×10^{-1}	3.3363	4.4536×10^{-1}

† Results corrected since the publication of Owens [14].

5 CONCLUSIONS

In the present paper we have presented a new BIM which is suitable for free surface boundary value problems involving creeping flows of a viscous fluid. The method has been applied to the problem of the planar extrusion of such a fluid. It has been pointed out in the Introduction that because the shape of the free surface is typically non-trivial and unknown a priori so-called singularity annihilation BIMs (see, for example, Kelmanson and Lonsdale [7]) are unusable, and that it is also impossible to employ singularity subtraction BIMs in such a way that the transformed variables have the required regularity properties throughout the flow domain.

The new BIM proposed in this article builds upon an existing idea of Hansen and Kelmanson [10], [11] which uses the change of variable (37) and integration by parts to perform the boundary integral along paths that pass through a singularity where $\partial \omega / \partial n$ may not be integrable but p is. We do the same. The essential difference between our formulation and that of Hansen and Kelmanson is in the treatment of the integration of $\partial p / \partial s G_j$ ($j = 1, 2$) in subintervals $\partial \Omega_i$ ($i = 1, 2, \ldots, n_1 + n_2 + n_3$) some distance from the singularity. It has been seen in Section 3.2.2 to be advantageous in such boundary elements to evaluate numerically the integral of $\partial p / \partial s G_j$ without performing an integration by parts. The result is a staggered boundary grid for the computation of p and the other flow variables.

The new BIM also has the advantage over the earlier work of Owens [14] in that we integrate around the true boundary of the flow domain whereas, due to the non-integrability of $\partial \omega / \partial n$ at the point of separation, Owens [14] skirted around this singular point following a circular arc, introducing an error, even though the radius of the arc centred at the separation point was small and the correct singular representations of the variables were employed along it.

The results shown in Section 4.2 for the extrudate swell ratio over a range of capillary numbers are in excellent agreement with others in the numerical literature [14], [17], [19], [25]. However, the extrudate swell ratio is a somewhat crude measure of the correctness of the solution. Therefore, in Section 4.3 we also post our calculated values for the separation angles α and leading order indices λ_1 appearing in the asymptotic form of the flow variables in the immediate neighbourhood of the separation point. Here there is a dire shortage of data in the literature with which we can compare our results although where theoretical [16], numerical [17] and experimental [26]–[29] results have been previously published agreement would appear to be convincing.

ACKNOWLEDGMENTS

The authors wish to thank Mark Kelmanson for pointing us to the references by Hansen and Kelmanson [10], [11] . We acknowledge the support of the Natural Sciences and Engineering Research Council of Canada (NSERC). Cette recherche a été financée par le Conseil de recherches en sciences naturelles et en génie du Canada (CRSNG).

REFERENCES

[1] Kelmanson, M.A., Modified integral equation solution of viscous flows near sharp corners. *Computers and Fluids*, **11**, pp. 307–324, 1983.

[2] Kelmanson, M.A., An integral equation method for the solution of singular slow flow problems. *J. Comput. Phys.*, **51**, pp. 139–158, 1983.

[3] Symm, G.T., Treatment of singularities in the solution of Laplace's equation by an integral equation method. Report NAC31, National Physical Laboratory, Teddington, UK, 1973.

[4] Ingham, D.B. & Kelmanson, M.A., A note on the comparison between BIE and FD techniques for solving elliptic BVPs with boundary singularities. *Commun. Appl. Numer. Methods*, **2**, pp. 189–193, 1986.

[5] Kelmanson, M.A., Solution of nonlinear elliptic equations with boundary singularities by an integral method. *J. Comput. Phys.*, **56**, pp. 244–258, 1984.

[6] Ingham, D.B. & Kelmanson, M.A., Solution of nonlinear elliptic equations with boundary singularities by an integral method. *Boundary Integral Equation Analyses of Singular, Potential and Biharmonic Problems*, eds C.A. Brebbia & S.A. Orszag, Springer Verlag, vol. 7 of *Lecture Notes in Engineering*, pp. 89–113, 1984.

[7] Kelmanson, M.A. & Lonsdale, B., Annihilation of boundary singularities via suitable Green's functions. *Computers Math. Applic.*, **29**, pp. 1–7, 1995.

[8] Moffatt, H.K., Viscous and resistive eddies near a sharp corner. *J. Fluid Mech.*, **18**, pp. 1–18, 1964.

[9] Karageorghis, A. & Fairweather, G., The method of fundamental solutions for the numerical solution of the biharmonic equation. *J. Comput. Phys.*, **69**, pp. 434–459, 1987.

[10] Hansen, E.B. & Kelmanson, M.A., Integral equation analysis of the driven-cavity boundary singularity. *Appl. Math. Lett.*, **5**, pp. 15–19, 1992.

[11] Hansen, E.B. & Kelmanson, M.A., An integral equation justification of the boundary conditions of the driven-cavity problem. *Computers and Fluids*, **23**, pp. 225–240, 1994.

[12] Xu, B., Some Problems in Slow Viscous Flow. PhD thesis, Danish Centre for Applied Mathematics and Mechanics, Lyngby, Denmark, 1985.

[13] Michael, D.H., The separation of a viscous liquid at a straight edge. *Mathematika*, **5**, pp. 82–84, 1958.

[14] Owens, R.G., The separation angle of the free surface of a viscous fluid at a straight edge. *J. Fluid Mech.*, **942**, pp. A50–1–A50–31, 2022.

[15] Anderson, D.M. & Davis, S.H., Two-fluid viscous flow in a corner. *J. Fluid Mech.*, **257**, pp. 1–31, 1993.

[16] Schultz, W.W. & Gervasio, C., A study of the singularity in the die-swell problem. *Q. J. Mech. Appl. Math.*, **43**, pp. 407–425, 1990.

[17] Salamon, T.R., Bornside, D.E., Armstrong, R.C. & Brown, R.A., The role of surface tension in the dominant balance in the die swell singularity. *Phys. Fluids*, **7**, pp. 2328–2344, 1995.

[18] Kelmanson, M.A., Boundary integral equation solution of viscous flows with free surfaces. *J. Engrg. Math.*, **17**, pp. 329–343, 1983.

[19] Mitsoulis, E., Georgiou, G.C. & Kountouriotis, Z., A study of various factors affecting Newtonian extrudate swell. *Computers and Fluids*, **57**, pp. 195–207, 2012.

[20] Ramalingam, S., Fiber Spinning and Rheology of Liquid-Crystalline Polymers. PhD thesis, Massachusetts Institute of Technology, Cambridge, MA, USA, 1994.

[21] Burda, P., On the FEM for the Navier–Stokes equations in domains with corner singularities. *Finite Element Methods, Superconvergence, Post-Processing and A Posteriori Estimates*, eds M. Křížek, P. Neittaanmäki & R. Stenberg, Marcel Dekker: New York, pp. 41–52, 1998.

[22] Lugt, H.J. & Schwiderski, E.W., Flows around dihedral angles, I: Eigenmotion analysis. *Proc. Roy. Soc. A*, **285**, pp. 382–399, 1965.

[23] Levenberg, K., A method for the solution of certain problems in least squares. *Quart. Appl. Math.*, **2**, pp. 164–168, 1944.

[24] Marquardt, D.W., An algorithm for least-squares estimation of nonlinear parameters. *SIAM J. Appl. Math.*, **11**, pp. 431–441, 1963.

[25] Georgiou, G.C. & Boudouvis, A.C., Converged solutions of the Newtonian extrudate-swell problem. *Int. J. Numer. Meth. Fluids*, **29**, pp. 363–371, 1999.

[26] Nickell, R.E., Tanner, R.I. & Caswell, B., The solution of viscous incompressible jet and free-surface flows using finite-element methods. *J. Fluid Mech.*, **65**, pp. 189–206, 1974.

[27] Tanner, R.I., Separation of viscous jets using boundary element methods. *Proceedings of the 9th Australasian Fluid Mechanics Conference*, University of Auckland School of Engineering: Auckland, New Zealand, pp. 247–250, 1986.

[28] Tanner, R.I., Lam, H. & Bush, M.B., The separation of viscous jets. *Phys. Fluids*, **28**, pp. 23–25, 1985.

[29] Batchelor, J., Berry, J.P. & Horsfall, F., Die swell in elastic and viscous fluids. *Polymer*, **14**, pp. 297–299, 1973.

[30] Batchelor, G.K., *An Introduction to Fluid Dynamics*. Cambridge University Press: Cambridge, UK, 1967.

[31] Longuet-Higgins, M.S., Mass transport in water waves. *Proc. R. Soc. Lond. A*, **245**, pp. 535–581, 1953.

[32] Sturges, L.D., Die swell: The separation of the free surface. *J. Non-Newtonian Fluid Mech.*, **6**, pp. 155–159, 1979.

[33] Shampine, L., Vectorized adaptive quadrature in MATLAB. *J. Comput. Appl. Math.*, **211**, pp. 131–140, 2008.

STOKES EQUATION SOLUTION USING THE LOCALIZED METHOD OF FUNDAMENTAL SOLUTIONS WITH A GLOBAL BASIS

JURAJ MUŽÍK & FILIP CIGÁŇ
Department of Geotechnics, Faculty of Civil Engineering, University of Zilina, Slovakia

ABSTRACT
The paper focuses on deriving a local variant of the method of fundamental solutions (MFS) for the case of Stokes flow. Compared to the global and local basis variants, the local with global basis one leads to a sparse characteristic matrix as in fully localized variants but with a narrower system of equations and thus makes the solution of especially large-scale problems more efficient. It is also essential to keep the condition number of the characteristic matrix within reasonable bounds and remove the solution dependency on fictitious sources. A combination of MFS and finite collocation approach was used for the localization with a globally defined Stokeslet fundamental solution. The results of the particular local variant were compared on several examples, and the dependence of the solution on the density of the point network and the dimensions of the stencil used were also tested in the paper.
Keywords: method of fundamental solutions, biharmonic equation, Stoke's flow.

1 INTRODUCTION
The linear case within the broader Navier–Stokes equations framework, which describes fluid motion, is called Stokes flow. In the context of former conditions, one can assume that the fluid is considered incompressible, and the flow velocity is very slow.

There are various approaches to solving the Stokes flow problem numerically. One such approach, the vorticity-stream function formulation, leverages the relationship in two dimensions between vorticity and the Laplacian of the stream function [1]. By eliminating vorticity, one obtains the biharmonic equation for the stream function, and the Stokes flow is solved in the form of the biharmonic equation.

Unlike traditional methods, reliant on predefined meshes, meshless numerical methods use scattered data points to approximate solutions in a continuous domain. These methods have gained popularity due to their effectiveness in handling complex geometries, adaptive refinement, and straightforward implementation. Over the past two decades, several meshless methods have emerged, including the Trefftz-like approaches represented by the method of fundamental solutions (MFS) [2]–[5], singular boundary method (SBM) [6] and boundary knots method (BKM) [7]. SBM, like MFS, employs the PDE's fundamental solution as an interpolation function, but it faces challenges related to singularities in the interpolation matrix. Conversely, the boundary knot method uses the general PDE solution as an interpolation function, resulting in regular diagonal terms. However, when dealing with a general PDE solution, the characteristic matrix may encounter issues related to its ill-conditioned character. Finding suitable general solutions for Laplace and biharmonic equations can be complex, and finding the proper technique to evaluate origin intensity factors in the case of SBM is also challenging. One of the possibilities for overcoming the mentioned issues is to use the fundamental solutions defined globally with the singular sources placed outside the application domain as in MFS and use them as a base for all local domains.

WIT Transactions on Engineering Sciences, Vol 135, © 2023 WIT Press
www.witpress.com, ISSN 1743-3533 (on-line)
doi:10.2495/BE460121

Recent attention has turned towards localized variations of the Trefftz-like method, aiming to enhance matrix conditioning. These variants adopt different localization techniques, primarily centred on applying the appropriate method within limited neighbourhoods of specific points. The side effect of the localized technique is the lowering of the fundamental solution source position influence on the final solution.

The MFS localized forms [3], [4] adopt mainly the local subdomain concept using the internal nodes with the fictitious boundary to evaluate the solution in the solution centre; the present method forms the local model using the convex hull of the local domain boundary nodes and the global sources divided regularly around the global domain. On the other hand, a localized numerical solution governed by a biharmonic equation (stream function form of the Stokes equation) is somewhat tricky because of the need to impose the two boundary conditions. A biharmonic problem is solved using the localized method of fundamental solutions presented in Fan et al. [5]. However, this formulation [5] uses the concept of a support domain with internal nodes and the problematic part caused by the imposition of the Neumann boundary conditions results in the overdetermined linear system with velocities in the direction of local model domains – the present formulation results in the regular linear system with the velocities as the part of the solution.

The localization principle is universally applicable, especially for methods emphasizing boundary points like the singular boundary method (SBM) or method of fundamental solutions (MFS). This principle defines a local solution region based solely on boundary points [8]. It mainly benefits the application of MFS, leading to the development of the local method of fundamental solutions (LMFS). In LMFS, MFS is used for local PDE solutions within overlapping subdomains of interior points, and a global sparse system of linear equations is assembled to determine the unknown values of the area.

The initial sections of this article present the description of the LMFS principle, along with their application to solving the two-dimensional Stokes flow problem. Subsequent sections present the results of solving the lid-driven cavity problem.

2 BASIC FLOW EQUATIONS

When the influence of inertial forces is much weaker than viscous forces and pressure gradient, one can label the flow as Stokes flow. Then, it is possible to describe 2D Stokes flow using the momentum (1) and continuity (2) as follows:

$$\frac{\partial p}{\partial x_i} - \Delta u_i = 0, \tag{1}$$

$$\frac{\partial u_i}{\partial x_i} = 0. \tag{2}$$

In these equations, u_i stands for the ith velocity component, and p represents pressure. Expressing the momentum and continuity equations relying on the concepts of vorticity ω and stream function Ψ (for more detail, see Fan et al. [9]) provides an alternative formulation of the Stokes equations as follows:

$$\Delta \omega = 0, \tag{3}$$

$$\Delta \Psi = -\omega. \tag{4}$$

Cancelling the vorticity term from eqn (4) results in the biharmonic form of the Stokes equation.

$$\Delta^2 \Psi = 0. \tag{5}$$

For boundary value problems governed by biharmonic PDE, the two types of boundary conditions should be specified simultaneously along the boundary Γ.

$$\Psi(\mathbf{x}) = 0 \qquad \mathbf{x} \in \Gamma, \tag{6}$$

$$\frac{\partial \Psi(\mathbf{x})}{\partial x_i} = (-1)^{(i+1)} u_{i0}. \tag{7}$$

In this context, u_{i0} represents the prescribed velocity components at the boundary Γ.

3 IMPLEMENTATION OF THE LOCALIZED MFS

The study aimed to solve eqn (5) while considering boundary conditions (6) and (7) within a two-dimensional domain Ω and its boundary Γ. We achieved this by creating a set of internal and boundary points that covered the entire global computational domain [10]. Near each internal node, we defined a small local domain (see Fig. 1). Within these domains, we applied the method of the fundamental solutions to solve the biharmonic equation, considering the unknown values at the local domain boundary, which are marked as 'solution centres'. The specified global boundary condition is applied if a solution centre is on the global boundary (6) and (7).

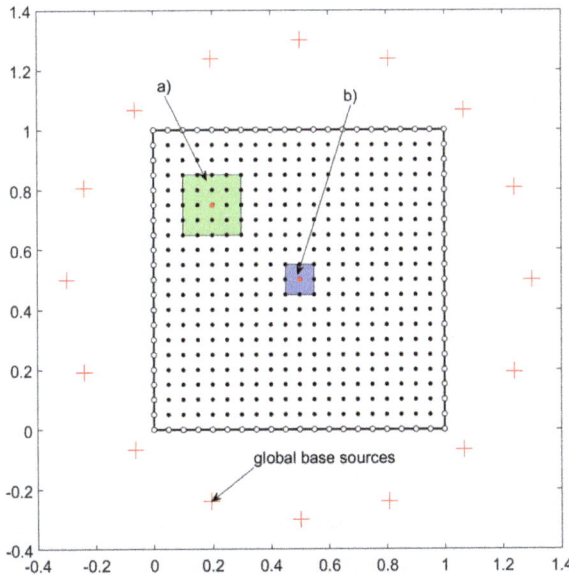

Figure 1: Regular local subdomains. (a) Eight solution centres; and (b) Sixteen solution centres on the local boundary.

For solving eqn (5) within each local domain Ω Ωk, we utilized the method of the fundamental solutions (MFS). In the evaluation of the stream function ψ and $\partial\psi/\partial\mathbf{n}$, we employed fundamental solutions of Laplace and the biharmonic operator [11] in the form:

$$\Psi(\mathbf{x}_i) = \sum_{j=1}^{n} G_{\mathrm{B}}(r_{ij})\alpha_j + G_L(r_{ij})\beta_j, \tag{8}$$

$$\frac{\partial \Psi(\mathbf{x}_i)}{\partial \mathbf{n}} = \sum_{j=1}^{n} \frac{\partial G_{\mathrm{B}}(r_{ij})}{\partial \mathbf{n}}\alpha_j + \frac{\partial G_L(r_{ij})}{\partial \mathbf{n}}\beta_j. \tag{9}$$

In these equations, n is the number of solution centres, and G_L and G_B were the fundamental solutions for Laplace and the biharmonic equation, respectively, which are defined as:

$$G_B(r_{ij}) = -\frac{1}{8\pi} r_{ij}^2 \ln(r_{ij}), \tag{10}$$

$$G_L(r_{ij}) = -\frac{1}{2\pi} \ln(r_{ij}), \tag{11}$$

where $r_{ij} = \|\mathbf{x}_i - \mathbf{x}_i^f\|$. To formulate the MFS solution within the local domain, we set up a system of linear equations:

$$\sum_{j=1}^{2n} A_{ij} a_j = d_i, i = 1, \ldots, 2n. \tag{12}$$

This system's structure depended on the local domain's number of solution centres n. Matrix $\mathbf{A}$ and vectors $\mathbf{a}$ and $\mathbf{d}$ took the following form:

$$\mathbf{A} = \begin{bmatrix} G_B(r_{cj}) & G_L(r_{cj}) \\ \frac{\partial G_B(r_{cj})}{\partial \mathbf{n}} & \frac{\partial G_L(r_{cj})}{\partial \mathbf{n}} \end{bmatrix}; \quad \mathbf{a} = \begin{Bmatrix} \alpha_j \\ \beta_j \end{Bmatrix}; \quad \mathbf{d} = \begin{Bmatrix} \Psi_j \\ \frac{\partial \Psi_j}{\partial \mathbf{n}} \end{Bmatrix}. \tag{13}$$

The values a_j in eqn (12) represented interpolation coefficients α and β for the given local domain. The vector d_i on the right side of eqn (12) contained the unknown values of Ψ and $\partial \Psi / \partial \mathbf{n}$ at the solution centres on the local domain's boundary. The values of Ψ and $\partial \Psi / \partial \mathbf{n}$ at the local domain's central point could be evaluated as follows:

$$\Psi_i(\mathbf{x}_c) = \sum_{j=1}^{n} G_B(r_{cj}) \alpha_j + G_L(r_{cj}) \beta_j, \tag{14}$$

$$\frac{\partial \Psi_i(\mathbf{x}_c)}{\partial \mathbf{n}} = \sum_{j=1}^{n} \frac{\partial G_B(r_{cj})}{\partial \mathbf{n}} \alpha_j + \frac{\partial G_L(r_{cj})}{\partial \mathbf{n}} \beta_j. \tag{15}$$

Here, r_{jc} is the radial distance between the central point x_c and solution centre j. Eqn (12) could be expressed in matrix notation as:

$$\mathbf{d} = \mathbf{G}_c \mathbf{a}. \tag{16}$$

From eqn (12), we could express the vector $\mathbf{a}$ as:

$$\mathbf{a} = [\mathbf{A}]^{-1} \mathbf{d}. \tag{17}$$

If we substituted this expression into eqn (13), we obtained:

$$\mathbf{b} = \mathbf{G}_c [\mathbf{A}]^{-1} \mathbf{d} = \mathbf{W}_c \mathbf{d}. \tag{18}$$

In this equation, $\mathbf{W}_c$ represented a stencil weight vector [12] that could be used to construct the global system of equations. This system, characterized by sparsity, could be defined as:

$$\Psi_k - \sum_{j=1}^{n} W_j^k \Psi_j - \sum_{j=n+1}^{2n} W_j^k \frac{\partial \Psi_j}{\partial x_i} n_{ij} = \sum_{l=1}^{m} W_l^k \Psi_{0l}, \tag{19}$$

$$\frac{\partial \Psi_k}{\partial x_i} - \sum_{j=1}^{n} \frac{\partial w_j^k}{\partial x_i} \Psi_j - \sum_{j=n+1}^{2n} \frac{\partial w_j^k}{\partial x_i} \frac{\partial \Psi_j}{\partial x_i} n_{ij} = \sum_{l=m+1}^{2m} W_l^k B_{im}. \tag{20}$$

In this representation, N was the number of internal points, n was the number of solution centres, m referred to the number of boundary centres in the kth local domain. This process resulted in forming a system of $3N$ sparse linear equations, which could be solved to obtain the stream function values and velocities at internal nodes.

4 STOKES FLOW – NUMERICAL EXAMPLE

The numerical example is a well-known one used to study Stokes flow. The geometrical conditions are formed by a square box filled with a thick liquid. Three sides of this box are still, meaning they do not move (the walls on the sides and the bottom). However, the top side, like a lid, can move. One can see the model configuration in Fig. 2, where we also show how the edges and boundary conditions are set.

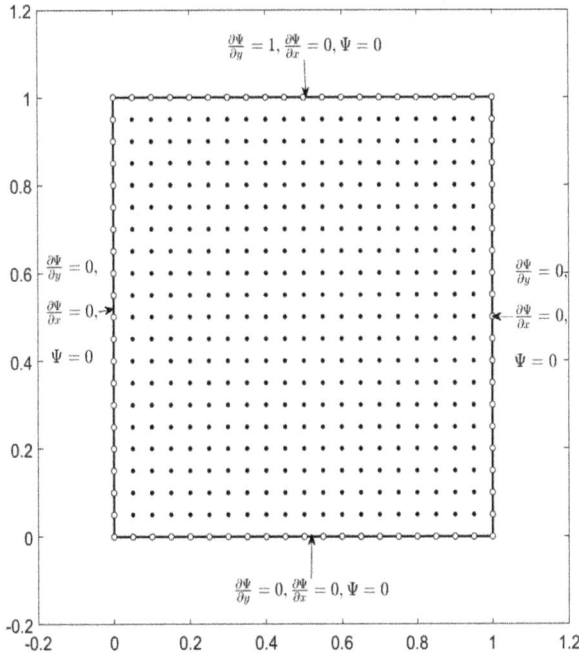

Figure 2: Lid-driven cavity, geometry, and boundary conditions.

In this case, we make the lid move at a steady speed, represented as $u_l = 1$. This motion also makes the liquid inside the box move, especially close to the walls. This creates swirling patterns and flows inside the box because the liquid sticks to the walls.

This numerical experiment is often used as a test problem when they want to try out new computer simulations or experiments [13]. It is a good choice because it is simple, and we know exactly what should happen. This helps to understand how the liquid sticks to the walls, how swirling happens, and when the flow separates from the walls.

In the present study the solution is obtained using three sets of points: one with a grid of 21×21 points, another with 41×41 points, and the last with 81×81 points. Fig. 3 shows the results as lines and arrows representing how the liquid moves at different places in the square cavity.

However, there is no exact solution for this problem in a closed form. So, we compare our results to what other researchers have found using computer codes, like the work in Botella and Peyret [13] and Mužík and Bulko [14]. Table 1 compares the highest and lowest values of scalar stream function ψ'' with what they found in Botella and Peyret [13]. Fig. 4 shows the velocity profiles compared with Song et al. [15].

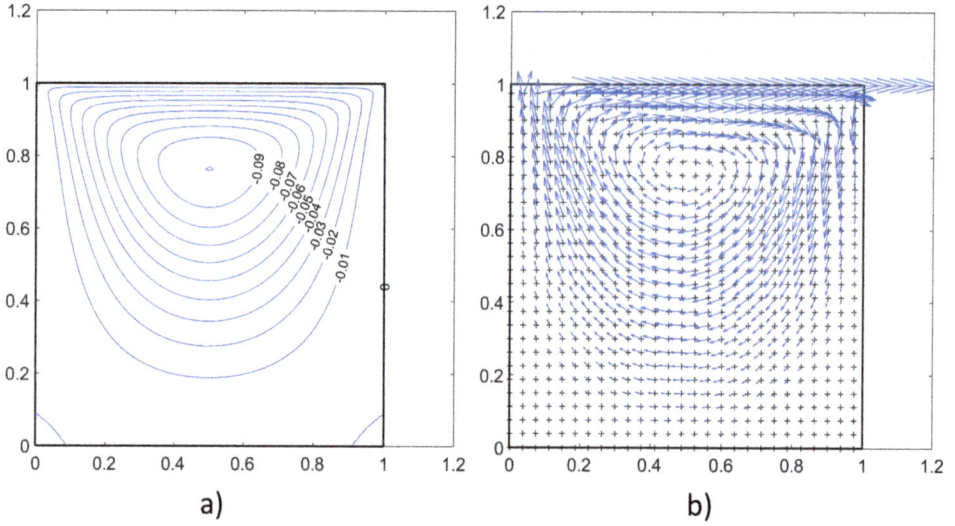

Figure 3:   Lid-driven cavity. (a) Contours of ψ; and (b) Velocity vectors.

Table 1:   Lid-driven cavity, comparison of ψmin and ψmax.

Mesh	$\psi_{\min}$	$\psi_{\max}$
21×21	-9.9788×10^{-2}	0
41×41	-9.9961×10^{-2}	1.7403×10^{-6}
81×81	-1.0007×10^{-1}	2.2261×10^{-6}
Botella and Peyret [13]	-1.0007×10^{-1}	2.2276×10^{-6}

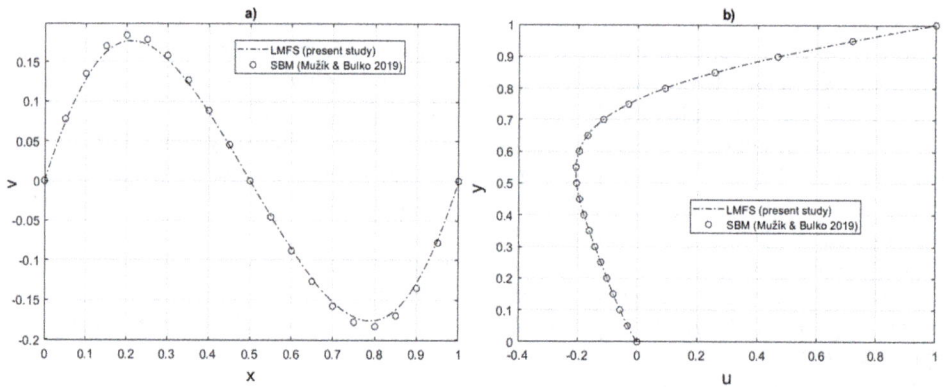

Figure 4:   (a) Velocity profile (u) on the centre line at $x = 0.5$; and (b) Velocity profile (v) on the centre line at $y = 0.5$.

5 CONCLUSIONS

This paper introduces a localized method of fundamental solutions for solving the problems related to Stokes flow. The advantage of this localized method of fundamental solutions is that it effectively manages the condition number of the global characteristic matrix, ensuring it remains reasonably well-conditioned. This improvement eliminates one of the primary shortcomings associated with this method. As a result, the overall matrix becomes sparse, maintains good conditioning, and can be solved using standard software tools.

As we continue to advance this method, a natural progression will be to expand its application to three-dimensional tasks or problems that involve non-linear behaviour.

ACKNOWLEDGEMENT

This contribution is the result of the project funded by the Scientific Grant Agency of Slovak Republic (VEGA) No. 1-0879-21.

REFERENCES

[1] Dean, E.J., Glowinski, R. & Pironneau, O., Iterative solution of the stream function vorticity formulation of the Stokes problem, applications to the numerical simulation of incompressible viscous flow. *Comput. Methods Appl. Mech. Eng.,* **87**, pp. 117–155, 1991.

[2] Golberg, M., The method of fundamental solutions for Poisson's equations. *Engineering Analysis with Boundary Elements,* **16**, pp. 205–213, 1995.

[3] Qu, W., Fan, C., Gu, Y. & Wang, F., Analysis of three-dimensional interior acoustic fields by using the localized method of fundamental solutions. Applied Mathematical Modelling, **76**, pp. 122–132, 2019.

[4] Gu, Y., Fan, C.M., Qu, W. & Wang. F., Localized method of fundamental solutions for large-scale modelling of three-dimensional anisotropic heat conduction problems: Theory and Matlab code. Computers Structures, **220**, pp. 144–155, 2019.

[5] Fan, C.M., Huang, Y.K., Chen, C.S. & Kuo, S.R., Localized method of fundamental solutions for solving two-dimensional Laplace and biharmonic equations. Engineering Analysis with Boundary Elements, **101**, pp. 188–197, 2019.

[6] Chen, W., Fu, Z. & Wei, X., Potential problems by singular boundary method satisfying moment condition. *CMES Comput. Model. Eng. Sci.,* **54**(1), pp. 65–85, 2009.

[7] Hon, Y.C. & Chen, W., Boundary knot method for 2D and 3D Helmholtz and convection–diffusion problems under complicated geometry. *Int. J. Numer. Methods Eng.,* **56**, pp. 31–48, 2003.

[8] Kovářík, K., Mužík, J., Bulko, R. & Sitányiová, D., Local singular boundary method for two-dimensional steady and unsteady potential flow. *Engineering Analysis with Boundary Elements*, **108**, pp. 68–78, 2019.

[9] Fan, C.M., Huang, Y.K., Li, P.W. & Lee, Y.T., Numerical solutions of two-dimensional Stokes flows by the boundary knot method. *CMES Comput. Model. Eng. Sci.,* **105**, pp. 491–515, 2015.

[10] Stevens, D. & Power, H., The radial basis function finite collocation approach for capturing sharp fronts in time dependent advection problems. *J. Comput. Phys.,* **298**, pp. 423–445, 2015.

[11] Xiong, J., Wen, J. & Liu, Y.C., Localized boundary knot method for solving two-dimensional Laplace and bi-harmonic equations. *Mathematic*, **8**, pp. 12–18, 2020.

[12]   Stevens, D., Power, H., Meng, C.Y., Howard, D. & Cliffe, K.A., An alternative local collocation strategy for high-convergence meshless PDE solutions, using radial basis functions. *J. Comput. Phys.*, **294**, pp. 52–75, 2013.

[13]   Botella, O. & Peyret, R., Benchmark spectral results on the lid-driven cavity flow. *Comput. and Fluids*, **27**(4), pp. 421–433, 1998.

[14]   Mužík, J. & Bulko, R., Multidomain singular boundary method for 2D laminar viscous flow. *Boundary Elements and Other Mesh Reduction Methods XLI*, pp. 31–41, 2019.

[15]   Song, L., Gu, Y. & Fan, C.M., Generalized finite difference method for solving stationary 2D and 3D Stokes equations with a mixed boundary condition. *Comput. Math. Appl.*, **80**, pp. 26–43, 2020.

BARNES–HUT/MULTIPOLE FAST ALGORITHM IN THE LAGRANGIAN VORTEX METHOD

ALEXANDRA KOLGANOVA, ILIA MARCHEVSKY & EVGENIYA RYATINA
Bauman Moscow State Technical University, Russia

ABSTRACT

New modification of the fast algorithm based on the Barnes–Hut (BH) and fast multipole method (FMM) is developed for the problem of velocities calculation in vortex particle method. It provides a quasilinear computational complexity and allows for the accuracy flexible adjustment, similar to the classic Barnes–Hut method. Four schemes are developed with a different number of terms being held in multipole and local expansions. All the necessary formulae are presented, expressed in terms of operations with complex numbers. If extremely high accuracy is not required, the proposed algorithm is more efficient in comparison to the traditional FMM methods.

Keywords: fast algorithm, n-body problem, vortex method, the Barnes–Hut algorithm, multipoles.

1 INTRODUCTION

Vortex particle method [1], [2] is an efficient tool for solving problems of incompressible flow simulation around airfoils including fluid structure interaction problems. The main idea is to consider the vorticity $\vec{\Omega} = \mathrm{curl}\vec{V}$ as a primary computational variable. For the viscosity effect simulation, we use the viscous vortex domains (VVD) method, based on the diffusive velocity approach, that seems to be the most efficient [3]. Thus, the Navier–Stokes equations in 2D case for incompressible flow can be written down in vorticity form:

$$\frac{\partial \vec{\Omega}}{\partial t} + \nabla \times (\vec{\Omega} \times (\vec{V} + \vec{W})) = \vec{0}, \tag{1}$$

where $\vec{V}$ is flow velocity, $\nabla \cdot \vec{V} = 0$; $\vec{\Omega} = \Omega \vec{k}$; $\vec{k}$ is unit vector orthogonal to the flow plane; $\vec{W} = -\nu \dfrac{\nabla \Omega}{\Omega}$ is the so-called diffusive velocity, proportional to the kinematic viscosity coefficient ν.

Eqn (1) can be considered as the transfer equation for vorticity that moves with the velocity $(\vec{V} + \vec{W})$. Vorticity field in the flow domain is simulated by a set of N vortex particles – Rankine or Lamb vortices, characterized by positions in the flow domain $\vec{r}_j$, circulations Γ_j that remain constant in time, $j = 1, \ldots, N$, and small radius ε, equal for all the vortices.

The velocity field can be reconstructed using the Biot–Savart law. If there are no airfoils in the flow domain, it takes the form

$$\vec{V}(\vec{r}) = \vec{V}_\infty + \int_S \frac{\vec{k} \times (\vec{r} - \vec{\xi})}{2\pi |\vec{r} - \vec{\xi}|^2} \Omega(\vec{\xi}) dS_\xi = \vec{V}_\infty + \sum_{j=1}^{N} \frac{\Gamma_j}{2\pi} \frac{\vec{k} \times (\vec{r} - \vec{r}_j)}{\max\{|\vec{r} - \vec{r}_j|^2, \varepsilon^2\}}, \tag{2}$$

here $\vec{V}_\infty$ is the incident flow velocity; Rankine vortex model is considered.

WIT Transactions on Engineering Sciences, Vol 135, © 2023 WIT Press
www.witpress.com, ISSN 1743-3533 (on-line)
doi:10.2495/BE460131

The vortex convective velocities calculation $\vec{V}(\vec{r}_i)$, $i = 1,\ldots,N$, according to eqn (2) is the most time-consuming operation in the vortex particle method algorithm. Being performed directly, its computational complexity is proportional to N^2; this problem is similar to the gravitational N-body problem. Taking into account that in vortex method the number of particles can reach millions [4], the only efficient way to reduce computational complexity is approximate fast methods implementation, which have quasilinear complexity. In order to choose optimal fast method, it is necessary to take into account specific features of the problem, the required accuracy, consistency of particular fast method with implementation of other operations in the whole algorithm, its scalability (for parallel computers of different architectures), etc. Note, that the considered problem differs significantly from the classical N-body problem: firstly, the kernel function in the integrand in eqn (2) decreases proportionally to distance, instead of squared distance, and secondly, vortex particles circulations can be positive or negative while body masses are always positive.

The tree-based Barnes–Hut method is the first fast method, suggested in 1986 for such problems [5], and it still remains popular. It has quasilinear computational complexity $O(N \log N)$ instead of squared one $O(N^2)$. Its adaptation for velocities calculation in vortex particle method is suggested in Dynnikova [6]. However, there are several issues: it is almost always less efficient in comparison to the fast multipole method (FMM) [8]; its adaptation for other operations in vortex method algorithm, e.g., the boundary integral equation solution, is not very efficient. Note also, that it can not be generalized for 3D flows simulation. The FMM algorithm, suggested slightly later in 1987, is also tree-based, but the tree traversal is performed in quite different way in comparison to the previously mentioned Barnes–Hut method. It has linear $O(N)$ computational complexity. Note, that it is very efficient when the extremely large number of particles is considered or very high, up to machine precision is required. But it is not obvious, that it is preferable in all the cases.

For example, in Capuzzo-Dolcetta and Miocchi [10], the authors had compared two methods for the gravitational problem by varying the accuracy and number of particles, and stated that the Barnes–Hut algorithm is more efficient if N has an order of hundreds of thousands. However, the multipole approximation works good for the clusters of dense-placed particles. If the problem implies essential non-uniform distribution of particles, hybrid mesh-hierarchical methods are known [11] that combine fast mesh algorithms (known as P3M [12]) for areas with rarefied-placed particles and multipole approximation for clusters. The ideas of constructing hybrid FMM/BH algorithm are also known [13] where high-order expansion of the influence function is used for electrostatic potential calculation, but without its local expansions. Taking into account that in vortex particle method the extremely high accuracy is not required, and the number of particles in 2D problems can reach a few million, the combination of the FMM and BH methods can lead to the optimal solution for the considered problems. This idea can be considered as the development of one proposed in Dynnikova [6] from more general point of view.

In the present paper a new modification of the Barnes–Hut algorithm is proposed, that has hybrid nature and includes some ideas of the Fast Multipole Method. The suggested algorithm seems to be preferable for 2D flows simulation due to flexible accuracy adjusting and convenient tree structure.

2 THE BARNES–HUT/MULTIPOLE ALGORITHM

The suggested method belongs to the tree-codes. The tree is the hierarchical structure of rectangular cells. The root cell includes all the particles in the flow domain. Then it is divided across the longest side, and two resulting first-level cells are cropped precisely to the particles, that are contained in them. In contrast to the original Barnes–Hut method, tree cells dividing is performed not up to level at which each cell contains a single vortex, but the process is stopped on the specified tree level. The example of tree cells for the vortex wake after the bridge cross section is shown in Fig. 1. The vortex wake is simulated with vortex particles.

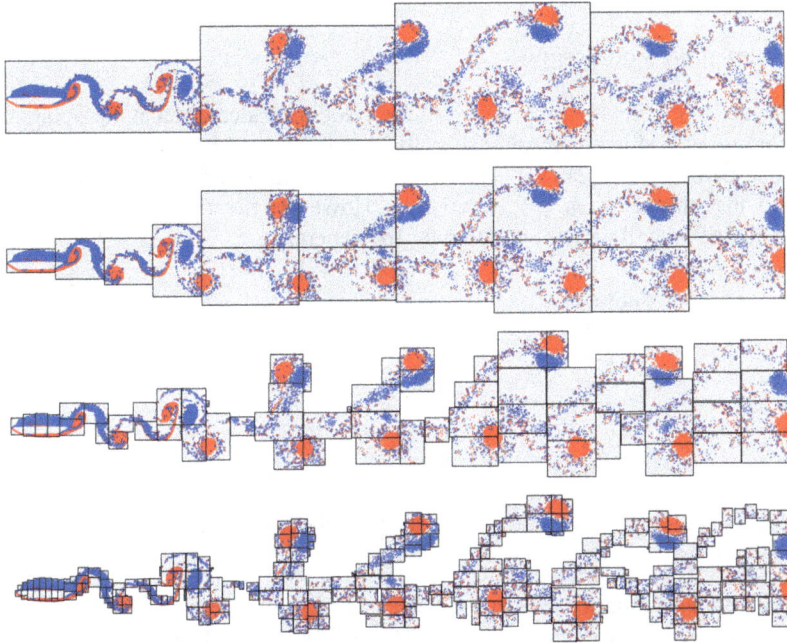

Figure 1: Tree cells example: 2nd, 4th, 6th and 8th level cells.

Note, that the above described algorithm of tree constructing can be called 'naive' and has the only advantage: it is adaptive, that is clearly seen in Fig. 1. However, it is recursive and poorly scalable; in parallel mode three construction for large values of N becomes very time-consuming procedure, takes up to half of time of the entire algorithm. Much more efficient way is to use non-recursive algorithm based on the fractal Morton's curve construction [7].

The main idea of the method suggested in the present paper, is that the influence of some far-placed cluster of particles onto another cluster, called 'control cluster' (more precisely, onto its centre), is calculated approximately using multipole expansion of the influence function. Then, the local expansion of the influence function is constructed in the control cluster, that is finally used for velocities computation at vortex particles positions. The velocity, induced by the vortex particles, placed in closely-placed clusters, is calculated directly according to the Biot–Savart law (2).

The method has two adjustable parameters: tree depth, that should be chosen in such a way to minimize computational complexity of the whole algorithm, and proximity parameter θ, that determines the ratio between the accuracy and computational complexity. We suppose that the cell p is far enough from another cell q if the distance between their centers $|\vec{\rho}|$ satisfies the proximity criterion:

$$|\vec{\rho}| \geq \frac{h_p + h_q + \varepsilon}{\theta}, \qquad (3)$$

where h_p and h_q are sums of width and length of the corresponding cells; vortices radius ε is added to numerator for the correct processing of cells containing a single particle, since the value of h for such cells is equal to zero.

2.1 Main idea of the modification

Let us consider firstly the model problem of the velocities calculation at the points, at which the particles are placed in a control cluster, that are influenced by vortex particles placed in an influence particle (Fig. 2). We assume, that the control and influence clusters are far enough, so that the criterion (3) is satisfied. However, the resulting formulae and the numerical algorithm will be written down for the general case.

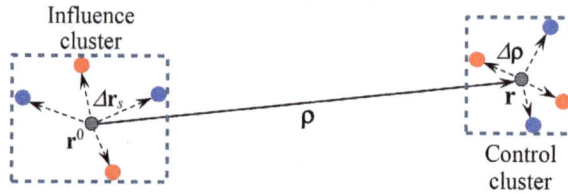

Figure 2: Scheme of two far-placed clusters interaction.

Here the following notations are introduced: $\vec{r}^0$ and $\vec{r}$ is the influence and control clusters centre positions, respectively; $\vec{\rho} = \vec{r} - \vec{r}^0$; $\vec{r}_s$ is the sth particle position in the influence cluster ($s = 1, \ldots, N_s$); $\Delta\vec{r}_s = \vec{r}_s - \vec{r}^0$; $\Delta\vec{\rho}$ is the vortex particle position in the control cluster with respect to its center.

The direct influence (hereinafter the 'influence' means the velocity, induced by the considered influence cluster at the corresponding point) at the control cluster centre can be calculated according to the Biot–Savart formula:

$$\vec{V}(\vec{r}) = \sum_{s=1}^{N_s} \Gamma_s \vec{k} \times (-\nabla_r G(\vec{r} - \vec{r}_s)), \qquad G(\vec{r}) = \frac{1}{2\pi} \ln\frac{|\vec{d}_*|}{|\vec{r}|},$$

here $\vec{d}_*$ is arbitrary constant non-zero vector.

Assuming that $|\Delta\vec{r}_s|$ is much smaller than $|\vec{\rho}|$, the influence function $\vec{V}(\vec{r})$ can be expanded into Taylor series with respect to the variables $|\Delta\vec{r}_s| / |\vec{\rho}|$:

$$\vec{V}(\vec{r}) = \sum_{s=1}^{N_s} \Gamma_s \vec{k} \times (-\nabla_r G((\vec{r} - \vec{r}^0) - \Delta\vec{r}_s)) = -\sum_{s=1}^{N_s} \Gamma_s \vec{k} \times (\nabla_\rho G(\vec{\rho} - \Delta\vec{r}_s))$$

$$\approx \vec{V}^m(\vec{\rho}) + \vec{V}^d(\vec{\rho}) + \vec{V}^q(\vec{\rho}) + \vec{V}^o(\vec{\rho}) + \vec{V}^h(\vec{\rho}), \qquad (4)$$

where $\vec{V}^m$, $\vec{V}^d$, $\vec{V}^q$, $\vec{V}^o$ and $\vec{V}^h$ are monopole, dipole, quadrupole, octupole and hexadecapole terms respectively, calculated as

$$\vec{V}^{(p)}(\vec{\rho}) = \frac{(-1)^{p+1}}{p!} \sum_{s=1}^{N_s} \Gamma_s \vec{k} \times (\nabla_\rho (\underbrace{\nabla_\rho \dots \nabla_\rho}_{p \ times} G(\vec{\rho}) \cdot \dots \cdot \underbrace{\Delta \vec{r}_s \otimes \dots \otimes \Delta \vec{r}_s}_{p \ times})), \tag{5}$$

where the operation '$\cdot$' means inner tensor product; '$\otimes$' is outer (Kronecker) tensor product; the superscript '(p)' corresponds to the p-th multipole term: $p=0$ for the monopole term, $p=1$ for the dipole term, etc. In fact, in the original Barnes — Hut algorithm monopole and dipole terms are taken into account because the centre of expansion is chosen at the centre of mass (the dipole moment in this case is equal to zero). Since vortex particles can have both positive and negative circulations, the vorticity centre, being understood similarly to the center of mass, can lie outside the cell or even at infinity (the last case takes place when total vorticity in the cell is equal to zero). Moreover, the accuracy of the expansion depends on position of expansion center. Thus, we choose the center of expansion at the geometrical cell center, that provides minimal a'priori error estimation. Another possible way is to consider vortices with positive and negative circulations separately, as it is suggested in Dynnikova [6]; in this case it is necessary to perform all the operations twice.

After series expansion of the influence function (4), all the multipole terms (5) should be factorized in order to distinguish the corresponding multipole moments that depend only on the sth cell parameters, hence they are calculated once for each cell. The general factorized formula is the following:

$$\vec{V}(\vec{r}) \approx \sum_p \vec{V}^{(p)}(\vec{\rho}) = \sum_p (\overline{\Theta}^{(p)}(\vec{\rho}) \cdot \dots \cdot \overline{m}^{(p)}(\Delta \vec{r}_s, \Gamma_s)),$$

where the multipole moments $\overline{m}^{(p)}$ are the pth rank tensors (scalar value $m^{(0)}$ is considered as the 0th rank tensor); the coefficients $\overline{\Theta}^{(p)}$ are the $(p+1)$th rank tensors equal to logarithmic potential multiple gradients up to constant factor $\pm 1/(2\pi)$:

$$\overline{\Theta}^{(p)}(\vec{\rho}) = (-1)^{p+1} \underbrace{\nabla_\rho \dots \nabla_\rho}_{(p+1) \ times} (\ln \frac{|\vec{d}_*|}{|\vec{\rho}|}), \quad p = 0, \dots, 4. \tag{6}$$

Here the notation $\overline{(\cdot)}$ is used for a tensor of arbitrary rank.

In the framework of traditional Barnes–Hut algorithm, a single particle is contained in each leaf cell, whereas the consideration of multiple particles per leaf cell requires much smaller tree (in term of tree depth), and therefore can be more efficient [9]. The optimal tree depth, that provides minimal computational complexity, depends on number of vortex particles as well as on geometrical shape of the vortex wake. Note, that the tree depth almost does not effect the accuracy of the velocities computation.

Thus, for vortex particles velocities calculation in the suggested fast method modification, the local expansions of the multipole terms in eqn (4) should be performed with respect to small distance $|\Delta \vec{\rho}|$ (Fig. 2). As a result, the velocity is calculated as follows:

$$\vec{V}(\vec{r} + \Delta \vec{\rho}) \approx \sum_p ((\overline{\Theta}^{(p)}(\vec{\rho}) + \nabla_\rho \overline{\Theta}^{(p)}(\vec{\rho}) \cdot \Delta \vec{\rho} + \dots) \cdot \dots \cdot \overline{m}^{(p)}(\Delta \vec{r}_s, \Gamma_s)).$$

This expression can be transformed and written down in suitable for practical computations form by grouping the terms containing multipliers with the same $\Delta\vec{\rho}$ degree (we denote $\Delta\vec{\rho}^{\otimes 2} = \Delta\vec{\rho} \otimes \Delta\vec{\rho}$, etc.):

$$\vec{V}(\vec{r} + \Delta\vec{\rho}) \approx \frac{\vec{k}}{2\pi} \times (\vec{E}_0(\vec{\rho}) + \hat{\vec{E}}_1(\vec{\rho}) \cdot \Delta\vec{\rho} + \frac{1}{2}\vec{\vec{E}}_2(\vec{\rho}) \cdot \cdot \Delta\vec{\rho}^{\otimes 2} + \frac{1}{6}\vec{E}_3(\vec{\rho}) \cdots \Delta\vec{\rho}^{\otimes 3}), \qquad (7)$$

where coefficients $\overline{\vec{E}}_q$ are tensors of the $(q+1)$th rank, which are accumulated for each control cluster over all influence clusters that are far enough from it according to the criterion (3). As it has been mentioned, the tensors $\overline{\vec{E}}_q$ are obtained as the coefficients in power (local) expansions of multipole terms, where the vector $\vec{E}_0$ corresponds to the sum of constant terms, the matrix $\hat{\vec{E}}$ – to linear terms, etc. Note, that number of terms in local expansion of multipole terms in (4) should be chosen consistently.

For example, if one considers only monopole term in eqn (4), there is no need to perform its local expansion at all: linear term of the local expansion has the same order of magnitude as the omitted dipole term. Then, if monopole and dipole terms are taken into account, the first one should be expanded up to linear term, while the second one can be considered as a constant, etc. However, in the table shown below, it is suggested to consider one additional multipole term (as constant, without local expansion). This approach allows to deal with influence clusters, which are much bigger than the control one; such situation is very common at the tree traversal. So, the following table shows the coefficients k_i, which are equal to 1 if the corresponding coefficient should be taken into account; otherwise they are equal to 0. The first row marked by I, corresponds to the coarsest approximation, where only two terms are taken into account without their local expansions; row II corresponds to more accurate version, etc.

	k_1	k_2	k_3	k_4
I	1	0	0	0
II	1	1	0	0
III	1	1	1	0
IV	1	1	1	1

For the most accurate scheme IV the tensor coefficients $\overline{\vec{E}}_q$ in local expansion are calculated using the following formulae:

$$\vec{E}_0(\vec{\rho}) = \vec{\vec{\Theta}}^m m^m + k_1 \hat{\vec{\Theta}}^d \cdot \vec{m}^d + \frac{k_2}{2!}\underline{\vec{\Theta}}^q \cdot \cdot \hat{\vec{m}}^q + \frac{k_3}{3!}\vec{\Theta}^o \cdots \underline{\vec{m}}^o + \frac{k_4}{4!}\tilde{\vec{\Theta}}^h \cdots \tilde{\vec{m}}^h,$$

$$\hat{\vec{E}}_1(\vec{\rho}) = -k_2 \hat{\vec{\Theta}}^d m^m - k_3 \underline{\vec{\Theta}}^q \cdot \vec{m}^d - \frac{k_4}{2!}\vec{\Theta}^o \cdot \cdot \hat{\vec{m}}^q, \qquad (8)$$

$$\vec{E}_2(\vec{\rho}) = k_3 \underline{\vec{\Theta}}^q m^m + k_4 \vec{\Theta}^o \cdot \vec{m}^d,$$

$$\vec{E}_3(\vec{\rho}) = -k_4 \vec{\Theta}^o m^m.$$

Eqn (7) is the final formula for approximate calculation of vortex particles velocities in control cluster, induced by far-placed influence clusters. However, calculations according to the presented formulae are extremely time-consuming due to operation on a lot of components of higher rank tensors. At the same time, all the tensors $\overline{\overline{m}}^{(p)}$, $\overline{\overline{\Theta}}^{(p)}$ and $\overline{\overline{E}}_q$ are fully symmetric (with respect to any pair of indices) and their convolutions over arbitrary pair of indices are equal to zero, so all of them (except monopole moment m^m which is scalar) are determined by only a pair of real numbers. Thus, in practice it is necessary to store and calculate only two components of each tensor.

2.2 Formulae for numerical calculations

Since we deal with 2D flow simulation, the vortex particle position in the flow domain $\vec{r}_s = \{x_s, y_s\}$ can be considered as a complex number $z_s = x_s + iy_s$, where i is the unit imaginary number. In further formulae we use the operations of complex numbers multiplication $z_1 \cdot z_2$ and raising to a power z^p.

Let us firstly consider the multipole moments calculation of tree leaf cells. The monopole moment is real scalar itself, so we still use the notation m^m. The tensor of any higher multipole moment $\overline{\overline{m}}^{(p)}$ is determined by two real numbers, denoted as $m_0^{(p)}$ and $m_1^{(p)}$, which can be considered as a real and imaginary parts of the complex number $m^{(p)} = m_0^{(p)} + im_1^{(p)}$. Thus, the multipole moments can be represented through vortex circulations Γ_s and complex numbers z_s that correspond to their positions $\Delta \vec{r}_s$, $s = 1,\ldots,N_s$, as follows:

$$m^{(p)} = \sum_{s=1}^{N_s} \Gamma_s z_s^p, \quad p \in \mathbb{N}. \tag{9}$$

Then the parent-cell parameters can be found by summation of children multipole moments (represented through complex numbers), being shifted to the centre of parent cell. The shifting scheme is shown in Fig. 3.

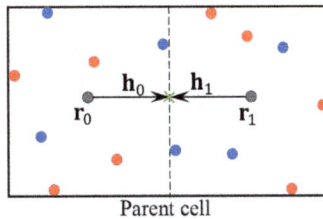

Figure 3: Multipole moments shifting.

The necessary formulae can be derived by calculation the multipole moments with respect to shifted point $(\vec{r}_i + \vec{h}_i)$, expressing the result through the same and lower-order unshifted moments. The resulting shifting rules, where shifting vector $\vec{h}$ is considered as a complex number $z_h = h_x + ih_y$, are the following:

$$m^{(p)}\big|_{\bar{h}} = \sum_{k=0}^{p} C_p^k m^{(p-k)} z_h^k,$$ (10)

where C_p^k is binomial coefficient. It should be noted, that all the presented formulae for the multipole moments and their shifting rules are similar to ones in the original FMM algorithm.

Tensor coefficients $\overset{=}{\Theta}{}^{(p)}$, that initially have been introduced according to eqn (6), now correspond to the complex numbers $\Theta^{(p)}$, and they are determined by only the distance vector between the clusters centers $\vec{\rho}$, represented as the complex number $z_\rho = \rho_x + i\rho_y$. Thus, the formulae, that are the most suitable in practice, have the following form:

$$\Theta^m = \frac{1}{|z_\rho|^2} z_\rho, \quad \Theta^{(p)} = \frac{p}{|z_\rho|^2}\left(\Theta^{(p-1)} \cdot z_\rho\right), \quad p \in \mathbb{N}.$$ (11)

In order to calculate complex analogues E_q of the tensor coefficients $\overset{=}{E}_q$ arising in local expansions introduced initially by eqn (8), it is necessary to summarize products of $\Theta^{(p)}$ and conjugate numbers $\overline{m}^{(p)}$:

$$E_0 = \Theta^m \cdot m^m + \Theta^d \cdot \overline{m}^d + \frac{1}{2!}\Theta^q \cdot \overline{m}^q + \frac{1}{3!}\Theta^o \cdot \overline{m}^o + \frac{1}{4!}\Theta^h \cdot \overline{m}^h,$$

$$E_1 = -\Theta^d \cdot m^m - \Theta^q \cdot \overline{m}^d - \frac{1}{2!}\Theta^o \cdot \overline{m}^q,$$ (12)

$$E_2 = \Theta^q \cdot m^m + \Theta^o \cdot \overline{m}^d,$$

$$E_3 = -\Theta^o \cdot m^m.$$

If one considers six multipole moments instead of five as we did, it is necessary to introduce additional coefficient E_4 in local expansion and add one more term in each expression for $E_0 \dots E_4$ in eqn (12), and similarly to higher number of multipole terms. So, all of the presented formulae, expressed through complex numbers allow applying them for arbitrary higher-order multipole moments and local expansions of corresponding terms.

Finally, in order to calculate the approximate velocity, instead of (7) the following formula for local expansions can be used:

$$\vec{V}(\vec{r} + \Delta\vec{\rho}) \approx \frac{\vec{k}}{2\pi} \times (\vec{U}^0 + \vec{U}^1 + \vec{U}^2 + \vec{U}^3),$$ (13)

where vectors $\vec{U}^0$, $\vec{U}^1$, $\vec{U}^2$ and $\vec{U}^3$ depend on the vector $\Delta\vec{\rho}$ and correspond to complex numbers U^q calculated as follows:

$$U^q = \frac{1}{q!}E_q \cdot \overline{z}_{\Delta\vec{\rho}}^q, \quad z_{\Delta\vec{\rho}} = \Delta\rho_x + i\Delta\rho_y, \quad q = 0,1,2,3.$$

2.3 Algorithm

The main steps of the suggested tree-based hybrid Barnes–Hut/multipole algorithm are the following:

1. Tree root formation, bounding all the vortex particles.
2. Hierarchical tree structure construction.
3. Calculation of the leaf cells parameters (complex analogues of multipole moments) according to eqn (9).
4. Upward tree traversal and parent-cells parameters calculation by summation shifted according to eqn (10) children multipole moments.
5. For each leaf cell:
   - downward tree traversal: if the current cell is far enough, i.e., the criterion (3) is satisfied, the local expansion coefficients (12) are accumulated; otherwise the downward traversal is continued or, if cell is a leaf, it is stored as a close cell list;
   - direct influence calculation induced by vortex particles, that are contained in close cells (using the Biot–Savart law);
   - summation of the last result with approximate influence calculated using eqn (13).

3 NUMERICAL EXPERIMENT

3.1 Accuracy investigation

The first numerical experiment is performed for the model problem of velocities calculation in square $h \times h$ control cluster induced by vortex particles contained in square $d \times d$ influence cluster placed at the distance $|\vec{\rho}|$ (Fig. 4).

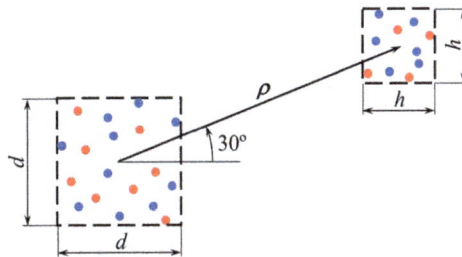

Figure 4: Model problem of two clusters interaction.

In order to estimate the accuracy of the developed method, we assume, that the influence cluster is always placed at the boundary of the close-zone, i.e., $\theta = (2d + 2h)/|\vec{\rho}|$ and we vary it by changing values h and $|\vec{\rho}|$. The relative error is calculated as

$$\delta V = \frac{\dfrac{1}{N}\sum_{i=1}^{N}|\vec{V}_i^{\,direct} - \vec{V}_i^{\,fast}|}{\max_i |\vec{V}_i^{\,direct}|},$$ (14)

where $\vec{V}_i^{\,direct}$ is the i th vortex velocity calculated directly by using the Biot–Savart law (eqn (2)); $\vec{V}_i^{\,fast}$ – using the proposed Barnes–Hut method modification. Fig. 5 shows the relative error δV dependency against the proximity parameter for the schemes I…IV that correspond to different number of terms accounted in series expansion.

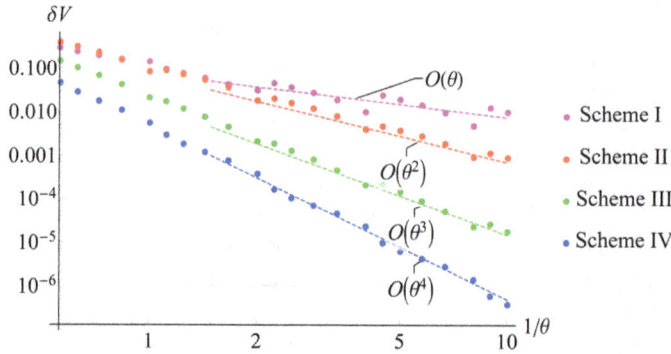

Figure 5:   The relative error of the velocities calculation against the inverse proximity parameter.

It is seen that the scheme I provides the first order of accuracy: the error decreases proportionally to θ, while the most accurate scheme IV provides the fourth order of accuracy (with respect to θ).

3.2 Complexity investigation

The method's computational complexity is examined on the problem of velocities calculation, induced by their mutual interaction, for large number of vortex particles: from $5 \cdot 10^4$ to $2 \cdot 10^6$. The number of multiplicative and division operations Q, performed on real numbers, is shown in Fig. 6 for plausible value of proximity criterion $\theta = 0.5$.

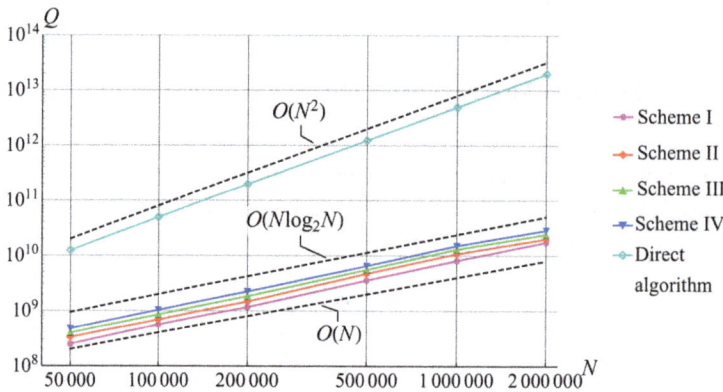

Figure 6:   Computational complexities of different fast method schemes against number of vortex particles.

It is seen that computational complexity of all the fast method modifications is quasilinear; the complexity of the most accurate scheme IV is twice as high in comparison to the scheme I, but it is one hundred times more accurate. In comparison to direct particle-to-particle (Biot–Savart) method, that has squared computational complexity, for the most time-consuming problem with 2 millions particles, the suggested fast method provides speedup up to 10^4 times.

3.3 Comparison with the FMM

In order to compare the proposed algorithm with original FMM method, we use in-house implementation of the FMM running on 6-core Intel i7-8700 CPU with OpenMP technology. We compare it to the most accurate version of the proposed fast method (scheme IV) also being run in parallel mode. Firstly, we consider two model problems of mutual interaction calculation for $N = 100000$ and $N = 500000$ vortex particles distributed uniformly in unit square. Calculation time for both methods is shown in Fig. 7, where 'B–H' means the proposed fast method modification. The labels under the 'columns' correspond to the relative error value, which is nearly the same for both methods.

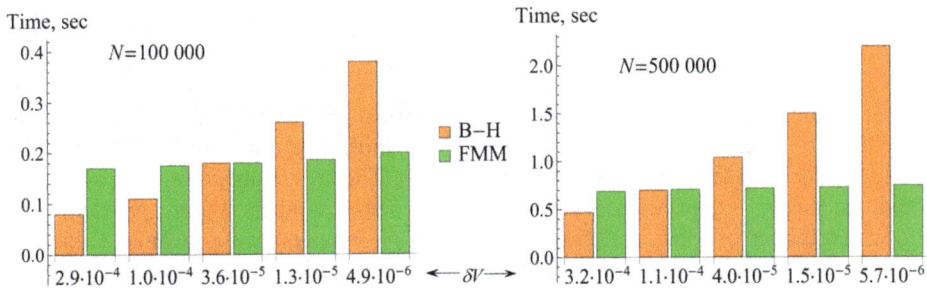

Figure 7:   Comparison of the Barnes–Hut method modification with original FMM algorithm for the uniformly distributed vortex particles; $N = 100000$ and $N = 500000$.

It is seen, that the FMM runtime does not change significantly with increasing the accuracy, so the method is especially efficient if the high accuracy is required. At the same time complexity of the proposed fast method depends essentially on the accuracy. It is slightly more efficient than FMM for low accuracy, but its efficiency reduces sharply with increasing both the accuracy and number of vortex particles.

The next numerical experiment is performed for similar problem, but now vortex particles form realistic vortex wake after the bridge cross-section (Fig. 1). Here we also consider two cases with different number of vortex particles – $N = 50000$ and $N = 250000$; calculation time is shown in Fig. 8.

So, that for the typical vortex particles distribution simulating some vortex wake, FMM runtime is two to three times higher in comparison to the proposed Barnes–Hut method modification at medium error level. Thus, for such problems the developed algorithm seems to be obviously preferable.

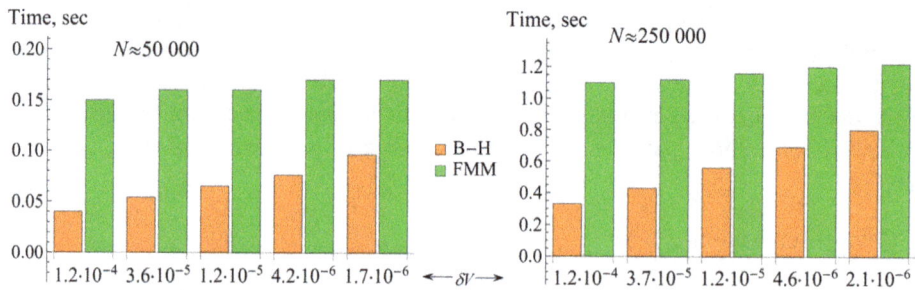

Figure 8: Comparison of the Barnes–Hut method modification with original FMM algorithm for the vortex wake after the bridge section, $N \approx 50000$ and $N \approx 250000$.

4 CONCLUSION

In the paper the Barnes–Hut algorithm modification for vortex particles velocities calculation in 2D vortex method is proposed. This method has quasilinear computational complexity and is based on multipole and local expansions of the influence function. It is tree-based algorithm and provides flexible accuracy adjusting in contrast to the original FMM method. Four schemes of different accuracy are proposed, that are distinguished by number of terms in corresponding expansions. Schemes have 1–4 orders of accuracy with respect to proximity parameter. The necessary formulae are derived for multipole moments calculation, their shifting rules and coefficients of the local expansions. The most suitable way for operations with all the tensors is their representation through complex numbers. For the problems with uniform distribution of vortex particles, the proposed algorithm takes approximately the same time as the FMM method at medium accuracy. For the velocities calculation of the particles, that form vortex wake after the bridge cross section, the proposed method is two to three times faster.

REFERENCES

[1]   Cottet, G.H. & Koumoutsakos, P.D., *Vortex Methods*: *Theory and Practice*, 2000.
[2]   Mimeau, C. & Mortazavi, I., A review of vortex methods and their applications: From creation to recent advances. *Fluids*, **6**(2), 68, 2021.
[3]   Guvernyuk, S.V. & Dynnikova, G.Y., Modeling the flow past an oscillating airfoil by the method of viscous vortex domains. *Fluid Dynamics*, **42**(1), pp. 1–11, 2007.
[4]   Kuzmina, K., Marchevsky, I., Soldatova I. & Izmailova, Y., On the scope of Lagrangian vortex methods for two-dimensional flow simulations and the POD technique application for data storing and analyzing. *Entropy*, **23**(1), 118, 2021.
[5]   Barnes, J. & Hut, P., A hierarchical $O(n \log n)$ force-calculation algorithm. *Nature*, **324**(4), pp. 446–449, 1986.
[6]   Dynnikova, G.Y., Fast technique for solving the N-body problem in flow simulation by vortex methods. *Computational Mathematics and Mathematical Physics*, **49**(8), pp. 1389–1396, 2009.
[7]   Karras T., Maximizing parallelism in the construction of BVHs, octrees, and k -d trees. *Proc. Intern. Conf. Eurographics/SIGGRAPH*, pp. 33–37, 2012.
[8]   Greengard, L. & Rokhlin, V., A fast algorithm for particle simulations. *Journal of Computational Physics*, **73**(2), pp. 325–348, 1987.

[9] Barnes, J., A modified tree code: Don't laugh; it runs. *Journal of Computational Physics*, **87**(1), pp. 161–170, 1990.

[10] Capuzzo-Dolcetta, R. & Miocchi, P., A comparison between fast multipole algorithm and tree-code to evaluate forces in 3-D. *Journal of Computational Physics*, **143**(1), pp. 28–48, 1997.

[11] Bode, P., Ostriker, J.P. & Xu G., The tree-particle-mesh N-body gravity solver. *Astrophysical Journal Supplementary*, **128**(2), pp. 561–570, 2000.

[12] Hockney, R.W. & Eastwood, J.W., *Computer Simulation using Particles*, CRC, 1988.

[13] Li, P., Johnston, H. & Krasny, R., A Cartesian treecode for screened Coulomb interactions. *Journal of Computational Physics*, **228**, pp. 3858–3868, 2009.

HIERARCHY OF NUMERICAL SCHEMES FOR THE BIE SOLUTION IN 2D FLOW SIMULATION USING VORTEX METHODS

YULIA IZMAILOVA, ILIA MARCHEVSKY & KSENIIA SOKOL
Bauman Moscow State Technical University, Russia

ABSTRACT

The problem of 2D incompressible flow simulation around airfoils with sharp edges and corner points is considered. The solution of the boundary integral equation with respect to vortex sheet intensity, arising in Lagrangian vortex method, has weak singularity that cannot be resolved correctly in the framework of the existing Galerkin-type numerical schemes. Known numerical schemes with piecewise-constant or piecewise-linear numerical solution representation provide solution reconstruction with high quality and the second order of accuracy only for piecewise-smooth bounded solutions. For singular solutions their order of accuracy goes down to the first. It is shown that wrong behaviour of numerical solution takes place only on the panels that adjust to the corner point. Modified numerical scheme is developed that is based on the Galerkin–Petrov approach and allows us to obtain integral characteristics of solution (the components of the added masses tensor) with the second order of accuracy. The scheme can be easily implemented in codes developed for flow simulations using the vortex particle method.

Keywords: vortex methods, boundary integral equation, unbounded solution, added mass tensor.

1 INTRODUCTION

Vortex methods of computational fluid mechanics [1] have rather narrow range of applicability: they are suitable for incompressible one-phase isothermal flows simulation. However, they are widely used in engineering applications [2], [3], especially connected with estimation of hydrodynamic loads acting on structures. Reviewing their advantages briefly, the following features should be pointed out:

- external and unbounded flows can be simulated with exact satisfaction of the perturbations decay boundary condition at infinity.
- efficient coupling schemes are developed that allow for solving fluid–structure interaction (FSI) problems with arbitrary translations, rotations, and deformations of the streamlined surface.
- the most part of computational resources is 'concentrated' in the part of the flow domain with non-zero vorticity; such area is usually rather compact.
- vortex methods belong to a class of particle-based methods, where Lagrangian particles are considered as vorticity carriers; it follows therefore that vortex methods provide rather small numerical diffusion.

Note that a wide list of investigations that are devoted to the problems connected with the vorticity field evolution simulation in the flow domain can be cited, and more or less efficient and accurate algorithms are developed that allow for simulation of vorticity convective transfer and its viscous diffusion. At the same time, the problems connected to the streamlined body (airfoil) simulation are investigated much poorly. In major part of papers and monographs only some general words can be found, mainly pointing out that the problem of the airfoil simulation is equivalent to the solving of some boundary integral equation (BIE) with respect to double layer potential density, vortex sheet intensity or vortex flux intensity.

WIT Transactions on Engineering Sciences, Vol 135, © 2023 WIT Press
www.witpress.com, ISSN 1743-3533 (on-line)
doi:10.2495/BE460141

And it is mentioned usually that the BIE can be solved using the same approaches as applied in the framework of the boundary element method. At the same time, the results of the accuracy investigation of the known numerical schemes that can be applied in 2D problems being solved by vortex methods, show that the error can be high enough, and it can be the most significant cause that bounds the accuracy of the flow simulation in the whole.

In the present paper, the approach is suggested that allows for increasing the accuracy of the BIE solution significantly for the airfoils with sharp edge or corner point, where the exact solution has the singularity.

2 THE GOVERNING EQUATION

We consider an approach to incompressible flow around an airfoil simulation, according to which it is necessary to solve the BIE with respect to vortex sheet intensity. For simplicity let us assume that the flow domain F is infinite and bounded only by the considered airfoil, and there is some known vorticity distribution $\Omega(\xi)$, $\xi \in F$, in the flow.

Then, the unknown vortex sheet intensity $\gamma(\mathbf{r})$ satisfies the vector boundary integral equation that arises from the no-slip condition satisfaction on the airfoil surface line:

$$\oint_K \frac{\mathbf{k} \times (\mathbf{r} - \xi)}{2\pi |\mathbf{r} - \xi|^2} \gamma(\xi) \, dl_\xi - \alpha(\mathbf{r}) \gamma(\mathbf{r}) \tau(\mathbf{r}) = \mathbf{f}(\mathbf{r}), \qquad \mathbf{r} \in K, \tag{1}$$

where K is the airfoil surface line; $\mathbf{k}$ is a unit vector orthogonal to the flow plane; $\alpha(\mathbf{r})$ is equal to the outer angle at the corresponding point divided by 2π (so, on smooth parts of the airfoil boundary $\alpha = 1/2$); $\tau(\mathbf{r})$ is tangent unit vector defined such as $\mathbf{n}(\mathbf{r}) \times \tau(\mathbf{r}) = \mathbf{k}$, $\mathbf{n}(\mathbf{r})$ is outer normal unit vector; the right-hand side is expressed as follows:

$$\mathbf{f}(\mathbf{r}) = -\left(\mathbf{V}_\infty + \int_F \frac{\mathbf{k} \times (\mathbf{r} - \xi)}{2\pi |\mathbf{r} - \xi|^2} \Omega(\xi) \, dS_\xi + \right.$$

$$\left. + \int_K \frac{\mathbf{k} \times (\mathbf{r} - \xi)}{2\pi |\mathbf{r} - \xi|^2} \gamma^{att}(\xi) \, dl_\xi + \int_K \frac{\mathbf{r} - \xi}{2\pi |\mathbf{r} - \xi|^2} q^{att}(\xi) \, dl_\xi - \alpha(\mathbf{r}) \mathbf{V}_K(\mathbf{r}) \right),$$

where $\mathbf{V}_\infty$ is incident velocity; $\gamma^{att}(\xi)$ and $q^{att}(\xi)$ are the so-called attached vortex and source sheet intensities, which are determined as tangent and normal components of the airfoil surface line velocity $\mathbf{V}_K(\mathbf{r})$, respectively:

$$\gamma^{att}(\xi) = \mathbf{V}_K(\xi) \cdot \tau(\xi), \qquad q^{att}(\xi) = \mathbf{V}_K(\xi) \cdot \mathbf{n}(\xi).$$

Note that in case of immovable airfoil the last three terms in the expression for the right-hand side are equal to zero; the assumption about the absence of the other surfaces in the flow is not essential; in order to take it into account, one should consider the system of the BIEs.

Traditionally, the following approach to numerical solution of eqn (1) is applied: firstly, it is projected onto the normal unit vector $\mathbf{n}(\mathbf{r})$, as the result, one obtains a singular BIE of the first kind with the Hilbert-type kernel $P_n(\mathbf{r}, \xi)$:

$$\oint_K \underbrace{-\frac{\boldsymbol{\tau}(\mathbf{r})\cdot(\mathbf{r}-\boldsymbol{\xi})}{2\pi|\mathbf{r}-\boldsymbol{\xi}|^2}}_{P_n(\mathbf{r},\boldsymbol{\xi})}\gamma(\boldsymbol{\xi})dl_\xi = \mathbf{f}(\mathbf{r})\cdot\mathbf{n}(\mathbf{r}), \qquad \mathbf{r}\in K. \tag{2}$$

For its solution special quadrature formulae are used (usually, similar to central quadratures rule) that allows for estimating the Cauchy principal value of the corresponding integrals. The family of such numerical schemes we call hereinafter N-scheme. Detailed investigation of such schemes with the proof of their convergence can be found in Lifanov [4] and Lifanov et al. [5].

Another approach leads to the so-called T-schemes and starts with the projection of eqn (1) onto tangent unit vector $\boldsymbol{\tau}(\mathbf{r})$ [6]. This means that it is necessary to solve the BIE of the second kind with bounded kernel (in case of smooth airfoil) or absolutely integrable kernel (in case of airfoils with sharp edges or corner points) $P_\tau(\mathbf{r},\boldsymbol{\xi})$:

$$\oint_K \underbrace{\frac{\mathbf{n}(\mathbf{r})\cdot(\mathbf{r}-\boldsymbol{\xi})}{2\pi|\mathbf{r}-\boldsymbol{\xi}|^2}}_{P_\tau(\mathbf{r},\boldsymbol{\xi})}\gamma(\boldsymbol{\xi})dl_\xi - \alpha(\mathbf{r})\gamma(\mathbf{r}) = \mathbf{f}(\mathbf{r})\cdot\boldsymbol{\tau}(\mathbf{r}), \qquad \mathbf{r}\in K. \tag{3}$$

For such an equation the family of more accurate numerical schemes can be developed; they are mostly based on the Galerkin approach.

3 SMOOTH AND NON-SMOOTH AIRFOILS

Let us firstly touch the case of smooth airfoil. The solution of the BIE here is smooth enough, and it can be reconstructed in principally without significant issues. We note only that the implementation of the above mentioned T-schemes may be non-trivial and requires rather accurate integrals computation that arise in Galerkin-type procedure [7].

Theoretic estimates, proven by numerical computations, show that the accuracy of numerical solution depends on the airfoil surface line discretization way. If it is replaced by a polygon consisting of straight panels, the order of accuracy (in L_1 norm) cannot be higher than the second: the first order of accuracy is achieved for piecewise-constant solution representation, the second order of accuracy – for piecewise-linear distribution of the vortex sheet intensity [8]. In much more complicated schemes, where the curvilinearity of the panels is taken into account explicitly [9], the third order of accuracy can be achieved at piecewise-quadratic numerical solution representation.

So, one can conclude that the problem of higher-order numerical schemes development is solved, more or less, for smooth airfoils.

At the same time, when one deals with the airfoil with sharp edge (i.e., Joukowsky wing) or corner points (i.e., rectangular or semicircular), the situation becomes more dramatic.

Let us firstly examine the case of Joukowsky wing airfoil (or some similar one). Initially, we note that the BIE (1), as well as its projections onto the normal and tangent directions (2) and (3), has infinite set of solutions. In order to pick out the unique solution, it is necessary, as a rule, to specify the total value of vorticity that is contained in the vortex sheet,

$$\oint_K \gamma(\boldsymbol{\xi})dl_\xi = \Gamma, \tag{4}$$

or, the same, to specify the velocity field circulation around the airfoil.

It is well-known that for wing airfoil there exists some particular value of circulation Γ^* that corresponds to the Chaplygin–Joukowsky condition, known also as Kutta condition, according to which the flow velocity remains finite at the sharp edge [10] (note that this condition is also applicable to a wider class of the airfoils with single corner points, e.g., the so-called generalized Joukowsky airfoils [11]). In this case, the solution of the BIE eqn (1) remains bounded and can have only jump discontinuity at the mentioned point (Fig. 1). For such solutions the above mentioned numerical schemes remain suitable, and they provide (sometimes with some obvious limitations) the same order of accuracy as for smooth solutions.

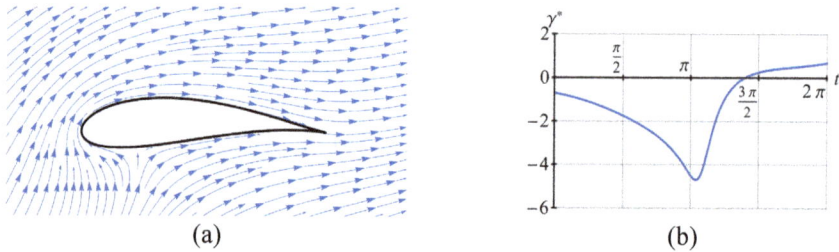

(a) (b)

Figure 1: (a) Flow around Joukowsky wing airfoil satisfying the Chaplygin–Joukowsky condition; and (b) Vortex sheet intensity with respect to a parameter that varies from 0 to 2π along the airfoil surface line.

However, the Chaplygin–Joukowsky condition corresponds to the steady-state flow regime that can be interesting in some cases, but such approach is not applicable to unsteady flows simulation, for example, when it is necessary to simulate transient regimes, etc. In this case, the circulation value Γ depends only on angular acceleration of the airfoil (i.e., if it moves plane-parallel or with constant angular velocity, then $\Gamma = 0$). Thereby, for unsteady flows around the airfoils with edges or corner points the solution of the BIE eqn (1) becomes singular in the corresponding point (Fig. 2), and its asymptotic behavior is known [4]:

$$\gamma(s) \sim s^{-\mu}, \qquad \mu = 1 - \frac{\pi}{\chi'} \tag{5}$$

where s is the distance to the edge; χ is the outer angle at the edge point. Consequently, for the classical Joukowsky airfoil with sharp edge there will be singularity of $s^{-1/2}$ type; for corner points with finite inner angle it has type $s^{-\mu}$, $0 < \mu < 1/2$.

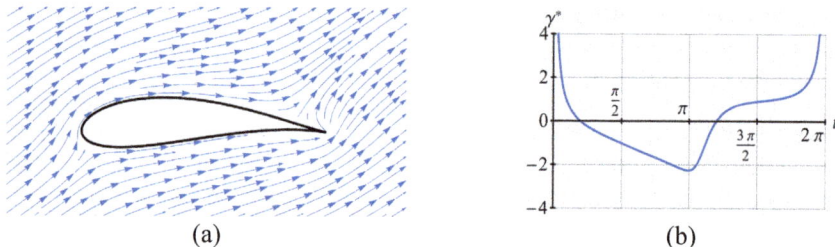

(a) (b)

Figure 2: (a) Flow around Joukowsky wing airfoil for $\Gamma = 0$; and (b) Vortex sheet intensity with respect to a parameter that varies from 0 to 2π along the airfoil surface line.

4 GALERKIN-TYPE NUMERICAL SCHEMES

In the above mentioned numerical schemes the Galerkin approach is implemented in order to derive the linear system that is the discrete analogue of the BIE.

The airfoil surface line K is replaced with the polygon consisting of the panels K_i, $i = 1, \ldots, N$. The solution is reconstructed as a combination of basis functions:

- for piecewise-constant solution representation

$$\gamma(\mathbf{r}) = \sum_{i=1}^{N} \gamma_i^0 \phi_i^0(\mathbf{r}), \qquad \mathbf{r} \in K, \tag{6}$$

- for piecewise-linear solution representation

$$\gamma(\mathbf{r}) = \sum_{i=1}^{N} \left(\gamma_i^0 \phi_i^0(\mathbf{r}) + \gamma_i^1 \phi_i^1(\mathbf{r}) \right), \qquad \mathbf{r} \in K, \tag{7}$$

where two families of basis functions are the following:

$$\phi_i^0(\mathbf{r}) = \begin{cases} 1, & \mathbf{r} \in K_i, \\ 0, & \mathbf{r} \notin K_i; \end{cases} \qquad \phi_i^1(\mathbf{r}) = \begin{cases} (\mathbf{r} - \mathbf{c}_i) \cdot \boldsymbol{\tau}_i / L_i, & \mathbf{r} \in K_i, \\ 0, & \mathbf{r} \notin K_i, \end{cases}$$

here $\mathbf{c}_i$ is the center of the panel K_i; L_i is its length.

In order to determine unknown coefficients γ_i^p values, the solution in the form (6) or (7) is substituted to the T-model (3), then the expression for the residual is multiplied by some projection function $\psi_i^p(\mathbf{r})$, and the result is assumed to be equal to zero. As the result, the algebraic linear system is obtained, from which the unknown coefficients are found.

The most obvious way is to choose the projection functions to be equal to basis ones; this corresponds to the classical Galerkin approach and for smooth airfoils (and, consequently, smooth and bounded solutions) can be considered as residual orthogonality condition in subspace of the L_2 functional space.

Such approach is implemented in Kuzmina and Marchevskii [7] where all necessary formulae for the integrals arising in the Galerkin procedure are presented. It works perfectly, providing the first and second orders of accuracy in case of smooth or at least bounded solutions in L_1 norm.

As an example, let us consider the problem of the added masses calculation. The added masses are widely used in various engineering applications in order to estimate the hydrodynamic forces acting the structure that moves in the flow with acceleration, i.e., vibrating in the flow. At the same time, according to Dynnikova [12] added mass tensor components also can be considered as integral characteristics of the vortex sheet intensity on the airfoil surface line that satisfies the BIE (1) or the equivalent eqns (2) and (3).

In order to estimate the added masses of the airfoil, three model problems should be considered: impulsive start of the airfoil in horizontal and vertical directions in still media with unit velocity $|\mathbf{V}_K| = 1$, and its impulsive start in rotational motion with unit angular velocity $\omega = 1$. The value of total circulation in additional condition (4) should be chosen as

$$\Gamma = -\oint_K \gamma^{att}(\xi) dl_\xi,$$

it is equal to zero except of rotational motion.

Then, the components of the added mass tensor for two-dimensional flow are calculated as follows [12]:

$$\lambda_{dx} = \oint_K \rho y \left(\gamma_d(\mathbf{r}) + \gamma_d^{att}(\mathbf{r}) \right) dl_r, \qquad \lambda_{dy} = -\oint_K \rho x \left(\gamma_d(\mathbf{r}) + \gamma_d^{att}(\mathbf{r}) \right) dl_r,$$

$$\lambda_{d\omega} = -\frac{1}{2} \oint_K \rho \left(x^2 + y^2 \right) \left(\gamma_d(\mathbf{r}) + \gamma_d^{att}(\mathbf{r}) \right) dl_r,$$

where x and y are the abscise and the ordinate of the point $\mathbf{r}$ at the airfoil boundary; index d has values x, y and ω and corresponds to the motion direction of the airfoil; $\gamma_d(\mathbf{r})$ and $\gamma_d^{att}(\mathbf{r})$ denote vortex sheets intensities in the airfoil motion in the d th direction; ρ is the flow density. The exact values are known for the added masses of the elliptic airfoil with semiaxes a and b [13]:

$$\lambda_{xx}^* = \rho \pi b^2, \quad \lambda_{yy}^* = \rho \pi a^2, \quad \lambda_{\omega\omega}^* = \frac{1}{8} \rho \pi \left(a^2 - b^2 \right)^2, \quad \lambda_{xy}^* = \lambda_{x\omega}^* = \lambda_{y\omega}^* = 0.$$

In Table 1 relative error of the added masses computation is given (maximal value for all diagonal components) for the elliptic airfoil with $5:1$ semiaxes ratio, which surface line is discretized into N panels of equal length. T^0 and T^1 denote numerical schemes with piecewise-constant and piecewise-linear solution representations. In brackets a'posteriori estimations for the order of accuracy are shown.

Table 1: Relative errors of the added mass tensor estimation for elliptical airfoil $5:1$ semiaxes ratio for different number of panels.

N	100	200	400	800	1600	3200
$\delta(T^0)$	$1.69\cdot10^{-2}$	$4.39\cdot10^{-3}$	$1.10\cdot10^{-3}$	$2.73\cdot10^{-4}$	$6.81\cdot10^{-5}$	$1.70\cdot10^{-5}$
Order	—	(1.94)	(2.00)	(2.01)	(2.01)	(2.00)
$\delta(T^1)$	$9.16\cdot10^{-3}$	$2.52\cdot10^{-3}$	$6.48\cdot10^{-4}$	$1.64\cdot10^{-4}$	$4.11\cdot10^{-5}$	$1.03\cdot10^{-5}$
Order	—	(1.86)	(1.96)	(1.98)	(1.99)	(2.00)

One can see that the second order of accuracy takes place for both numerical schemes; the scheme T^1 is 40% more accurate in comparison to T^0. Note that the order of accuracy for the T^0 scheme is higher than for vortex sheet intensity; this is due to the fact that the solution is smooth. For the T^1 scheme the orders of accuracy are equal to 2 in both cases.

Now let us examine these schemes for Joukowsky wing airfoil [13] (Fig. 3).

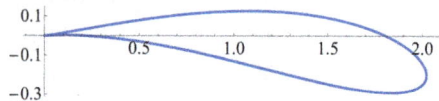

Figure 3: Joukowsky airfoil shape for the parameters $a=1$, $\eta=1.15$, $\alpha=\pi/30$.

The specific feature of this airfoil is that in horizontal motion at $\Gamma = 0$ in condition (4) the exact solution for vortex sheet intensity $\gamma(s)$, where s is arclength coordinate, is bounded and has jump discontinuity at the sharp edge (Fig. 4(a)).

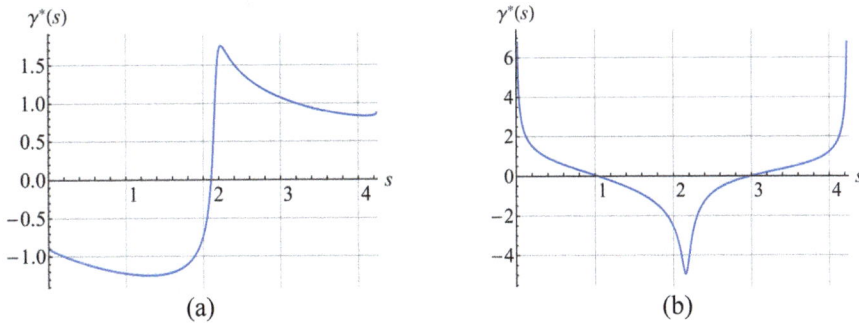

Figure 4:   Exact solution for the vortex sheet intensity $\gamma^*(s)$ for the Joukowsky wing airfoil in (a) horizontal; and (b) vertical motion. s is arc length measured along the airfoil surface line from the sharp edge.

The expressions for added mass tensor components for such airfoil could be found in Sedov [13].

In Table 2 maximal relative error of the added masses computation is given for λ_{xx}, $\lambda_{x\omega}$ and $\lambda_{y\omega}$ (the components that can be obtained after horizontal airfoil motion simulation) for the wing airfoil discretized into N panels of equal length. Again, in brackets a'posteriori estimations for the order of accuracy are shown.

Table 2:   Relative errors of the added masses λ_{xx}, $\lambda_{x\omega}$ and $\lambda_{y\omega}$ estimation for Joukowsky wing airfoil for different number of panels.

N	100	200	400	800	1600	3200
$\delta(T^0)$	$6.77 \cdot 10^{-3}$	$1.61 \cdot 10^{-3}$	$3.71 \cdot 10^{-4}$	$8.25 \cdot 10^{-5}$	$1.92 \cdot 10^{-5}$	$4.79 \cdot 10^{-6}$
Order	—	(2.07)	(2.12)	(2.17)	(2.10)	(2.00)
$\delta(T^1)$	$4.55 \cdot 10^{-3}$	$1.18 \cdot 10^{-3}$	$2.99 \cdot 10^{-4}$	$7.50 \cdot 10^{-5}$	$1.88 \cdot 10^{-5}$	$4.70 \cdot 10^{-6}$
Order	—	(1.95)	(1.98)	(1.99)	(2.00)	(2.00)

Here we see that both schemes again provide the second order of accuracy and their errors are nearly the same for fine discretization.

However, if we consider all the components of the added mass tensor, which calculation requires vertical and rotational airfoil motion simulation, where solution is unbounded near the sharp edge (Fig. 4(b)), we obtain much coarser result (Table 3).

The scheme T^1 is now approximately 10 time more accurate than T^0, but both schemes provide now not higher than the first order of accuracy due to not enough accuracy of solution reconstruction on the panels that are adjacent to the sharp edge (Fig. 5).

Table 3:   Maximal relative errors of the all the added masses estimation for Joukowsky wing airfoil for different number of panels.

N	100	200	400	800	1600	3200
$\delta(T^0)$	$2.51 \cdot 10^{-2}$	$1.29 \cdot 10^{-2}$	$6.88 \cdot 10^{-3}$	$3.73 \cdot 10^{-3}$	$2.03 \cdot 10^{-3}$	$1.11 \cdot 10^{-3}$
Order	—	(0.96)	(0.91)	(0.88)	(0.87)	(0.87)
$\delta(T^1)$	$4.55 \cdot 10^{-3}$	$1.65 \cdot 10^{-3}$	$7.40 \cdot 10^{-4}$	$3.51 \cdot 10^{-4}$	$1.71 \cdot 10^{-4}$	$8.49 \cdot 10^{-5}$
Order	—	(1.46)	(1.16)	(1.08)	(1.03)	(1.01)

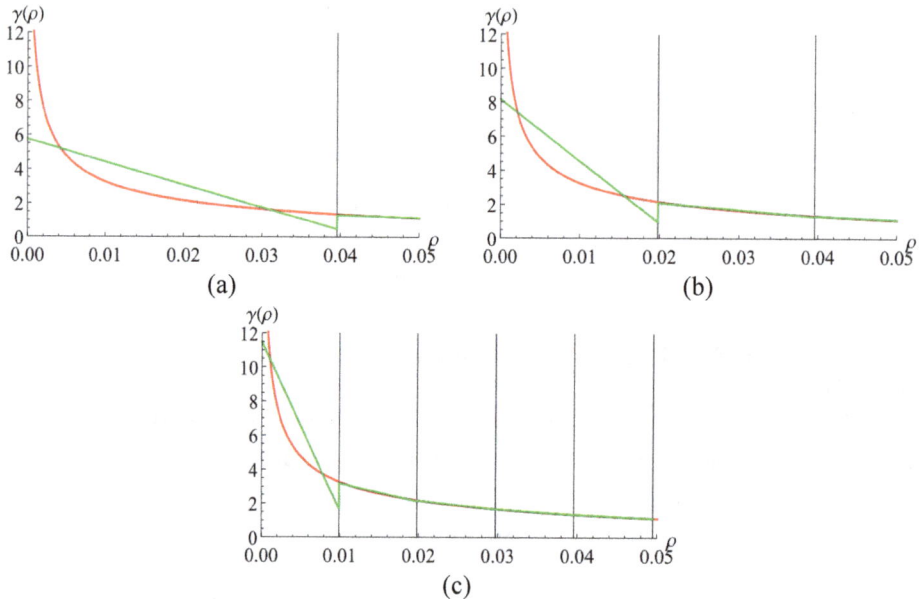

Figure 5:   Vortex sheet intensity against the distance to the sharp edge obtained by using the T^1 scheme with piecewise-constant solution representation (green line) and the exact solution (red curve); number of panels $N = 100$ (a); $N = 200$ (b); and $N = 400$ (c); vertical lines correspond to panel endings.

Such an issue cannot be overcome in the framework of the schemes T^0 and T^1 since the solution now belongs to L_1 functional space instead of space containing bounded smooth functions, or piecewise-smooth bounded functions with fixed point of jump discontinuity.

5 NUMERICAL SCHEME WITH SINGULAR BASIS FUNCTIONS

It is seen from Fig. 5 that the scheme T^1 is quite suitable over all the panels except those, which are adjacent to the sharp edge. This put us onto an idea of the modification of this scheme that allows for taking into account singularity of the solution. Let us suggest the following representation for the solution:

$$\gamma(\mathbf{r}) = \left(\gamma_1^0 \phi_1^0(\mathbf{r}) + \gamma_1^a \phi_1^a(\mathbf{r})\right) + \sum_{i=2}^{N-1}\left(\gamma_i^0 \phi_i^0(\mathbf{r}) + \gamma_i^1 \phi_i^1(\mathbf{r})\right) + \left(\gamma_N^0 \phi_N^0(\mathbf{r}) + \gamma_N^a \phi_N^a(\mathbf{r})\right), \quad \mathbf{r} \in K, \quad (8)$$

where on the first and N th panels the solution basis functions

$$\phi_1^a(\mathbf{r}) = \frac{L_1^\mu}{s_1(\mathbf{r})^\mu} - \frac{1}{1-\mu}, \qquad \overline{\phi}_N^a(\mathbf{r}) = \frac{L_N^\mu}{\left(L_N - s_N(\mathbf{r})\right)^\mu} - \frac{1}{1-\mu}$$

are introduced instead of linear ones; $s_i(\mathbf{r})$ means the distance from the beginning of the i th panel to the point $\mathbf{r}$. Such choice of basis functions is consistent with the asymptotic behavior of the exact solution (5); the constant $(1-\mu)^{-1}$ is added in order to provide their average value equal to zero.

However, the traditional Galerkin approach, according to which projection functions are equal to basis ones, now does not seem to be suitable. First of all, we note that it is applicable in principle only for $\mu < 1/2$; it means that we can consider only airfoils with corner points, but not with sharp edges, since in the last case the integrals over the first and the N th panels

$$\int_{K_1} \left(\phi_1^a(\mathbf{r})\right)^2 dl_r = \infty, \qquad \int_{K_N} \left(\overline{\phi}_N^a(\mathbf{r})\right)^2 dl_r = \infty,$$

i.e., they do not converge. This follows from the fact that the exact solution for the airfoil with sharp edge $(\mu = 1/2)$ does not belong to L_2 functional space, but belongs to L_1 space.

In order to develop universal algorithm, we use Petrov–Galerkin approach, according to which the set of projection functions differs from set of basis ones. Namely, we use now the same projection functions as in original $\mathcal{T}^1$ scheme: $\{\phi_i^0\}_{i=1}^N$ and $\{\phi_i^1\}_{i=1}^N$, i.e., piecewise-constant and piecewise-linear ones. Another advantage of such approach is connected to the computation of the coefficients of the resulting matrix: only $4N$ coefficients should be replaced by new values, while $4N(N-1)$ remain the same as in the matrix arising in the $\mathcal{T}^1$ scheme (note, that the right-hand side remains exactly the same).

The results of the BIE solution obtained by using the suggested scheme (let us denote it $\mathcal{T}_a^1$) for the above considered wing airfoil with corner point are shown in Fig. 6.

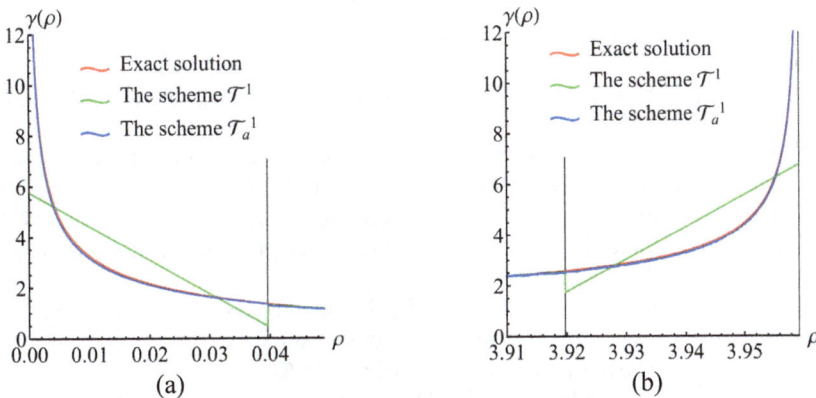

Figure 6:   Vortex sheet intensity over the first (a) and the N th (b) panels of the wing airfoil for the surface line discretization into $N = 100$ panels; vertical lines = panel endings.

It is clearly seen that the quality and accuracy of solution reconstruction on the first panel is now much higher than for original $\mathcal{T}^1$ scheme. Over the N th panel the exact and numerical solutions behavior is quite similar.

Let us now estimate the quality of the developed scheme $\mathcal{T}_a^1$ by estimating the added mass components tensor for the Joukowsky airfoil with sharp edge that was considered above (Fig. 3). The maximal relative error for all the added masses are shown in Table 4.

It is seen that the second order of accuracy is provided due to solution singularity accounting.

Table 4: Maximal relative errors of the all the added masses estimation for Joukowsky wing airfoil for different number of panels.

N	100	200	400	800	1600	3200
$\delta(\mathcal{T}_a^1)$	$4.59 \cdot 10^{-3}$	$1.19 \cdot 10^{-2}$	$3.01 \cdot 10^{-4}$	$7.56 \cdot 10^{-5}$	$1.89 \cdot 10^{-5}$	$4.74 \cdot 10^{-6}$
Order	—	(1.95)	(1.98)	(1.99)	(2.00)	(2.00)

6 CONCLUSION

The problem of the numerical solution of a boundary integral equation of the second kind arising in 2D vortex methods with respect to the intensity of the vortex sheet at the airfoil surface line is considered. In case of the airfoil with a corner point or a sharp edge, the exact solution of the integral equation becomes unbounded, which does not allow the boundary integral equation to be numerically solved with high accuracy using the Galerkin approach with bounded basis functions. A scheme based on the Petrov–Galerkin method is proposed as an original numerical scheme that allows one to reconstruct unbounded solutions. In this scheme, constant basis functions and unbounded basis functions with a known asymptotic behavior are introduced on the panels where the solution is unbounded, on the other panels constant and linear basis functions are used; on all panels, constant and linear functions are chosen as projection functions. The developed scheme allows one to improve the accuracy significantly.

REFERENCES

[1] Cottet, G.-H. & Koumoutsakos, P.D., *Vortex Methods: Theory and Practice*, CUP, 2000.

[2] Lewis, R.I., *Vortex Element Methods for Fluid Dynamic Analysis of Engineering Systems*, Cambridge, 1991.

[3] Branlard, E., *Wind Turbine Aerodynamics and Vorticity-Based Methods: Fundamentals and Recent Applications*, Springer, 2017.

[4] Lifanov, I.K., *Singular Integral Equations and Discrete Vortices*, Utrecht, 1996.

[5] Lifanov, I.K., Poltavsky, L.N. & Vainikko, G., *Hypersingular Equations and their Applications*, CRC Press, 2004.

[6] Kempka, S.N., Glass, M.W., Peery, J.S., Strickland, J.H. & Ingber, M.S., Accuracy considerations for implementing velocity boundary conditions in vorticity formulations. SANDIA report SAND96-0583, 1996.

[7] Kuzmina, K.S. & Marchevskii, I.K., On the calculation of the vortex sheet and point vortices effects at approximate solution of the boundary integral equation in 2D vortex methods of computational hydrodynamics. *Fluid Dyn.*, **54**, pp. 991–1001, 2019.

[8]   Kuzmina, K.S., Marchevskii, I.K., Moreva, V.S. & Ryatina E.P., Numerical scheme of the second order of accuracy for vortex methods for incompressible flow simulation around airfoils. *Russ. Aeronaut.*, **60**, pp. 398–405, 2017.

[9]   Marchevsky, I.K., Kuzmina, K.S. & Soldatova, I.A., Improved algorithm of boundary integral equation approximation in 2D vortex method for flow simulation around curvilinear airfoil. *Mathematics and Mathematical Modeling*, **6**, pp. 22–51, 2018.

[10] Tokaty, G.A., *A History and Philosophy of Fluid Mechanics*, New York, 1994.

[11] Glauert, H., A generalised type of Joukowski aerofoil. *NACA Reports and Memoranda*, **911**, 1924.

[12] Dynnikova, G.Y., Added mass in a model of a viscous incompressible fluid. *Dokl. Phys.*, **64**, pp. 397–400, 2019.

[13] Sedov, L.I., *Two-Dimensional Problems in Hydrodynamics and Aerodynamics*, Geneva, 1965.

OPEN-SOURCE PARALLEL CODES FOR 2D AND 3D FLOW SIMULATION BY LAGRANGIAN VORTEX METHODS

ILIA MARCHEVSKY, EVGENIYA RYATINA, GEORGY SHCHEGLOV & KSENIIA SOKOL
Bauman Moscow State Technical University, Russia

ABSTRACT

Meshless Lagrangian vortex methods that are characterized by considering vorticity as a primary computational variable are discussed, including their modern modifications for 2D and 3D flow simulation. Original mathematical models developed by the authors are described, that allow for significant improvement of the accuracy of the flow simulation around the airfoils/bodies. The hierarchy of numerical schemes based on the Galerkin approach is developed for numerical solution of the boundary integral equation. The quality of the surface mesh is not essential, rather high quality of the numerical solution can be achieved even for low-quality mesh consists of triangular cells with high aspect ratio. The open-source parallel codes (for CPU and GPU, using OpenMP, MPI and Nvidia CUDA technologies) are developed, that implement viscous vortex domains method and closed vortex loops method for 2D and 3D cases, respectively In 3D cases, the numerical scheme allows to satisfy the divergence-free condition for vorticity field (in 2D it is done trivially). The suggested methods can be applied for unsteady hydrodynamic load computation at rather low computational cost of the algorithm. The developed models and algorithms are suitable for numerical simulation in coupled problems, including for light movable bodies. Both weakly-coupled and strongly-coupled strategies are implemented, the last one requires several iterations; at each of them the boundary integral equation is solved. In addition to flow simulation and hydrodynamic load estimation, the suggested technique allows for added masses tensor calculation with high accuracy. Efficient fast method of quasilinear numerical complexity, both well-known and developed by the authors for the integral equation solution and vortex particles (that simulate the vorticity distribution in the flow domain) evolution simulation are discussed. A number of numerical examples are presented, being performed for validation of the developed mathematical models, numerical algorithms and parallel codes.

Keywords: incompressible flows simulation, open-source code, FSI problems, OpenMP, MPI, Nvidia CUDA.

1 INTRODUCTION

Vortex methods of computational fluid dynamics are based on considering the vorticity as a primary computational variable and they belong to a class of pure Lagrangian particle-based methods or have hybrid Lagrangian–Eulerian nature.

The basic idea of vortex methods, at least in the 2D case is connected with fundamental result, that have been established by Professor N. E. Zhukovsky in 1906: if one considers velocity field in inviscid incompressible flow around some airfoil, then this field would coincide exactly with the velocity field, that is induced by incident flow and the influence of vortex sheet of some intensity, placed on the surface line of the airfoil. Thus, the airfoil can be replaced with a vortex sheet, and for its intensity determination either analytic or numerical approaches are used. When a separate airfoil of a simple shape is considered in the unbounded flow domain, a conformal mappings technique can be applied (elliptic airfoil, semi-circular airfoil, Zhukovsky wing airfoil and some others). For more complicated shapes the problems are reduced to solving the boundary integral equation, which can be of Fredholm-type (with bounded kernel), singular (with Cauchy/Hilbert-type kernel, the integral is understood in the sense of its principal value) or hypersingular (in Hadamard sense). A

WIT Transactions on Engineering Sciences, Vol 135, © 2023 WIT Press
www.witpress.com, ISSN 1743-3533 (on-line)
doi:10.2495/BE460151

number of numerical schemes are developed, which provide a different level of accuracy and complexity. The most popular and most well-known schemes follow from solving of Laplacian/Helmholtz equation with respect to potential, with the boundary conditions of the second kind, i.e., the Neumann problem: the normal derivative of the potential should be equal to some given function on the airfoil surface. Considering the solution in the form of the double-layer potential, one can easily derive the corresponding hyper-singular integral equation of the first kind.

However, there is no need to find the double layer density potential Φ as such, instead of it vortex sheet intensity can be considered, that is mathematically equal to the surface gradient of Φ, multiplied by normal unit vector: $\vec{\gamma} = \vec{n} \times (\nabla \Phi)$. Applying the necessary tricks, the equation is reduced (in 2D case) to singular one, that is solved in most known implementations of vortex methods.

At the same time, the other approach was suggested in Kempka et al. [1], that allows for considering a Fredholm-type equation of the second kind with respect to vortex sheet intensity: instead of boundary condition for normal component of velocity (that is equal to the normal derivative of the potential in the framework of potential flow), the condition for the tangent component of velocity is considered. For smooth airfoils, the kernel of the resulting equation is bounded, for airfoils with corner points or sharp edges the kernel is unbounded, but the corresponding integrals are understood in traditional sense (as improper ones).

Note, that the briefly described mathematical model can be generalized and used not only for solving the problems for inviscid flow (described by Euler equations), but also for viscous flows simulation, i.e., the Navier–Stokes equation solution, both in 2D and 3D cases. In the 3D case, the idea remains the same: vortex sheet intensity, which is now not scalar, but vector lying in tangent plane, is considered as unknown, and a vectoral boundary integral equation (BIE) of the second kind is solved, that expresses the boundary condition for the tangent velocity component. Such an approach is more efficient than the well-known approach [2] for constructing BIE of the first kind that expresses the boundary condition for the normal velocity component for which numerical schemes such as the scheme of discrete vortex method and some others are developed (N-schemes).

For solving BIE of the second kind the family of numerical schemes, for 2D and 3D simulations, based on the briefly described approach and Galerkin projection method, is called 'T-schemes' ('T' means 'Tangent') [3]. Their usage in flows simulation seems to be essential: in the 2D case of viscous flows simulations, it is necessary to provide a high quality of velocity field reconstruction in the near-body area (in the boundary layer). Note, that for viscous flow vorticity flux is simulated from the body surface to the flow; according to Lighthill's approach, not attached, but a free vortex sheet is introduced, vorticity from it is discretized into a large number of small vortex particles, that move in the flow domain and form boundary layer and vortex wake. Stochastic and deterministic algorithms for simulation of such type are known: random walk method (RWM) [4], particle strength exchange (PSE) method [5], vorticity redistribution method (VRM) [6] and viscous vortex domains method (VVD) [7]. Namely, the VVD will be used in this work.

In 3D case there is no known purely Lagrangian methods; the review of known approaches can be found in Mimeau and Mortazavi [8]; let us focus only on vorticity generation procedure on the body surface. We suggest to use the vortex loops method, according to which vorticity in the flow domain is represented as closed structures: each loop is considered as closed vortex tube of small radius, which is generated on the body surface, and then moves in the flow. All vortex tubes have equal intensity (circulation), that makes it possible to simulate their merging, reconnections, etc. Initial positions of vortex loops coincide with level-set lines of double-layer potential density on the body surface, however, the results are much more accurate if the following strategy is used: firstly, the BIE is solved with respect to vortex sheet (using the T-scheme), and secondly, the potential density is reconstructed by using the least squares method. This two-steps scheme normally provides much better result in comparison to 'direct' hyper-singular BIE solution with respect to the potential density, especially for low-quality surface meshes.

2 BRIEF REVIEW OF EXISTING CODES

Despite the fact, that vortex methods have a long history [9], [10] dating back to the 1930s…1950s, up to now there are just a few software implementations, which are freely available for researchers, based on actual mathematical models and allows to use modern computational technologies (we do not take in mind the so-called 'panel methods', implemented in some codes, since they can be considered only as a very simple modification of vortex methods, that is suitable for some specific problems). Some codes have appeared in recent years; let us confine ourselves to listing them without detailed description of their capabilities.

The vvflow [11] (available from 2018) is based on the VVD method and allows to simulate for 2D flow. Coupled FSI problems can be considered for rigid elastically connected airfoils. For computations only OpenMP technology is used; so it can be run on multicore processors with shared memory; CUDA is not supported.

Codes Omega2D [12] and Omega3D [13] have appeared in 2018, they are based on VRM method. Only flow around immovable airfoils/bodies can be simulated; however, the codes are developed intensively, e.g., in 2022 the fast algorithm (of Barnes–Hut type) is implemented for vortex particle influence computation. Parallel computational technologies are not used 'explicitly', however, OpenMP and CUDA can be 'implicitly' used through the third-party libraries.

In 2019, the FLOWVPM code have appeared as a part of the FLOWUnsteady project [14]. 3D flows around bodies, fixed and movable, are considered. Vortex blobs are considered as vortex particles; note, that in this code fast multipole method (FMM) used for vortex particle influence computation, however parallel computational technologies are not widely used.

It is also known about the VXflow code [15], which is used actively by developers, however, it is not freely available. This code is for 2D problems; it is based on RWM approach [4] and the fast Fourier transform (FFT) technique for vortex particles interaction simulation. Algorithms for parallelization are based on OpenCL technology, so some part of computations can be transferred to a graphic card.

In all mentioned codes traditional approach is used for the boundary condition satisfaction: singular or hyper-singular BIE is solved, usually by using rather coarse numerical schemes.

3 THE VM2D AND VM3D CODES

Since the freely available codes now are based on outdated mathematical models and in most cases do not allow using modern high-performance computational technologies, the authors developed original implementations of vortex methods (VM) for 2D and 3D cases, based on T-schemes, VVD (in 2D) and closed vortex loops method (in 3D), and intensive use of capabilities of both multicore/multiprocessor computers and modern graphic cards. The source code of VM2D is open and freely available in GitHub public repository:

http://github.com/vortexmethods/VM2D (accessed on 26 Jun. 2023); the VM3D code will be available soon.

The general flowchart of both algorithms is shown in Fig. 1; there are no principal differences in blocks of VM2D and VM3D, however, most parts of specific algorithms differ significantly.

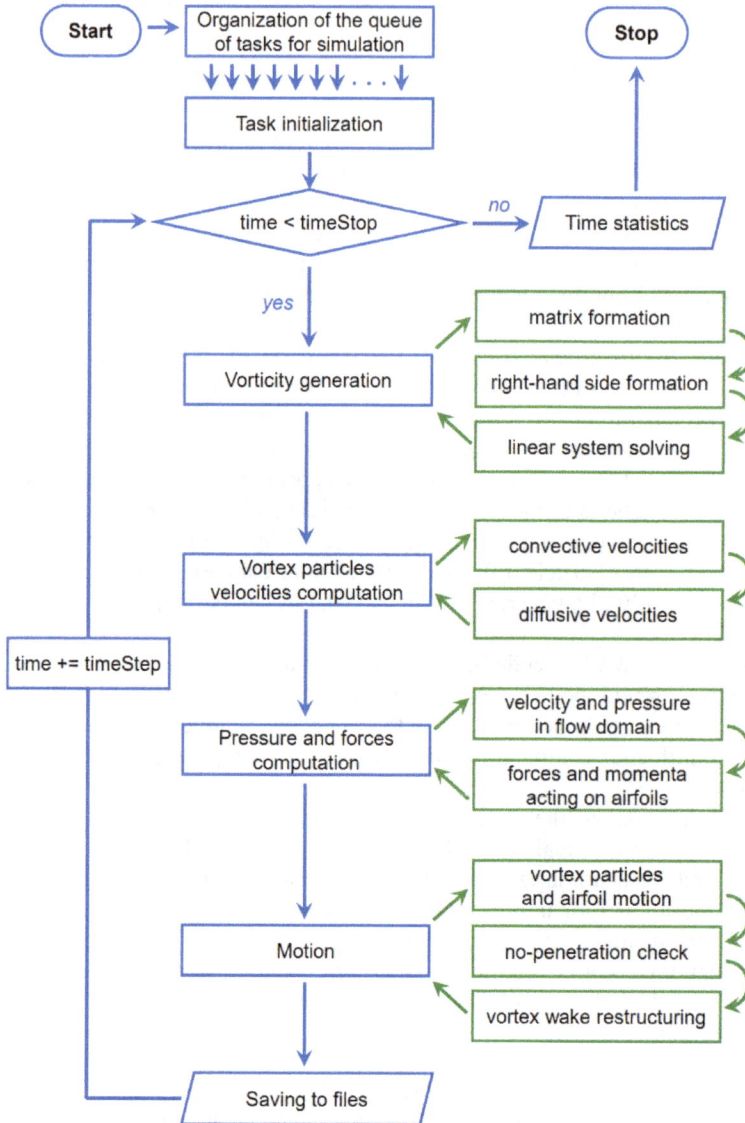

Figure 1: Flowchart of vortex method algorithm.

Codes are written in C++ language, cross-platform and have a modular structure. Parallel technologies OpenMP, MPI and Nvidia CUDA are used for performing computations on shared memory systems, distributed memory cluster systems and graphic accelerators, respectively.

To improve the efficiency of computations performed on CPU, at least for the most time-consuming computational blocks of the algorithm (particle velocities computation and BIE solving) it is reasonable to use special fast algorithms. Such an algorithm is developed as prototypes and applied for both mentioned problems: it is a hybrid Barnes–Hut/multipole algorithm, that has of quasilinear computational complexity [16], [17]. Its implementation is rather efficient, it allows to achieve speedup of about 1,800 times for the velocities computation of 2 million vortex particles in parallel mode (OpenMP, 18 cores). The time of BIE solving for elliptical airfoil discretized by $N = 3,200$ panels at one time step is reduced by 50 times in comparison with the Gaussian elimination algorithm for the corresponding system of linear equations.

However, in the current version of open codes only direct algorithms are implemented (fast algorithms will be available in the nearest future). It means that in order to reduce computation time it is necessary to use a lot of CPU cores or graphics accelerators. To estimate the efficiency of parallel implementation of VM2D code for the problem of flow simulation around a circle airfoil, which is discretized by $N = 2,000$ panels, is considered at Re = 3,000. Computations are performed on the cluster with 84 nodes, each is equipped with 28 cores (2xIntel Xeon E5-2690v4). The efficiency of parallel implementation is about 70% for fixed time-step with approximately 600 000 vortex particles in the vortex wake. It corresponds to 0.45% of sequential code, according to the Amdahl's law. Since the vortex method is a particle method, and all the particles can be processed independently, its parallel implementation adapted for calculations using graphics accelerators is extremely efficient: computational time of one time-step with 600,000 vortex particles by using a single GPU Tesla V100 is approximately the same as using 84 nodes, each with 2×14-core CPU.

4 NUMERICAL RESULTS

4.1 Unsteady hydrodynamic loads computation for impulsively started cylinder

In the VM2D code for the BIE solution the T-schemes with piecewise-constant and piecewise-linear solution representation are implemented for rectilinear panels that discretize airfoil surface line. In order to verify the developed T-schemes for the boundary integral equation solution the well-known test problem of flow simulation around an impulsively started circular cylinder is considered at the Reynolds number Re = 200 (Fig. 2).

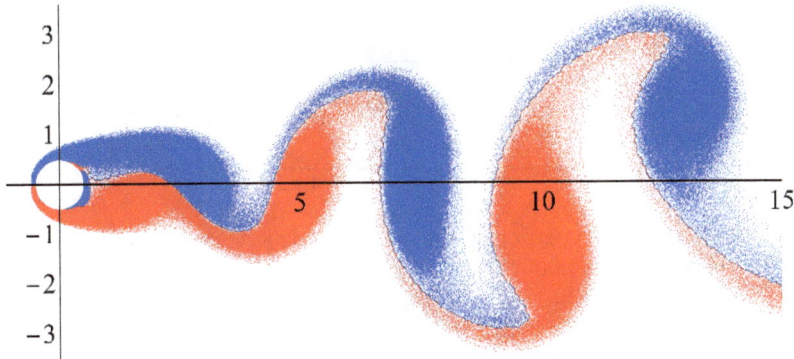

Figure 2: Vortex wake after circle cylinder.

Numerical experiments show that the usage of different schemes influences significantly the hydrodynamic loads acting on the airfoil. Values of unsteady drag and lift coefficients against time are shown in Fig. 3. It is seen that usage of the N-scheme provides rather high oscillation, so some special filtering procedure is required, while usage of the T-scheme provides much smaller amplitude of oscillations.

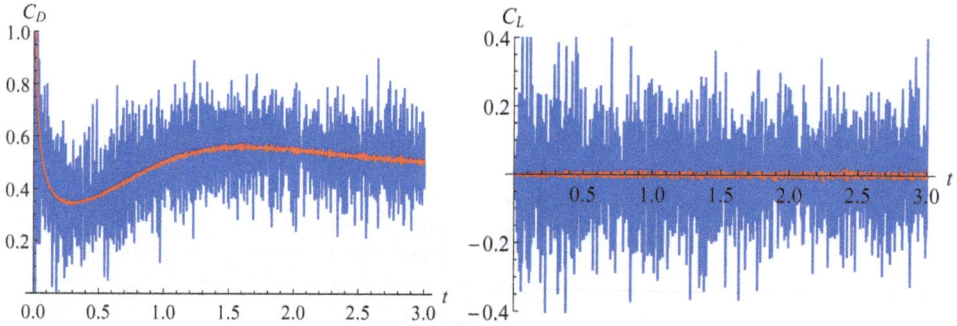

Figure 3:   Hydrodynamic loads oscillations for flow simulation around impulsively started cylinder at Re = 200: N-scheme (blue) and T-scheme (red).

Another numerical experiment was performed for a similar problem at Re = 3,000. The drag coefficient for first 5 seconds is shown in Fig. 4 in comparison with results obtained by other authors [18]–[24]. A good agreement is observed. Results obtained by using VM2D are presented without any averaging.

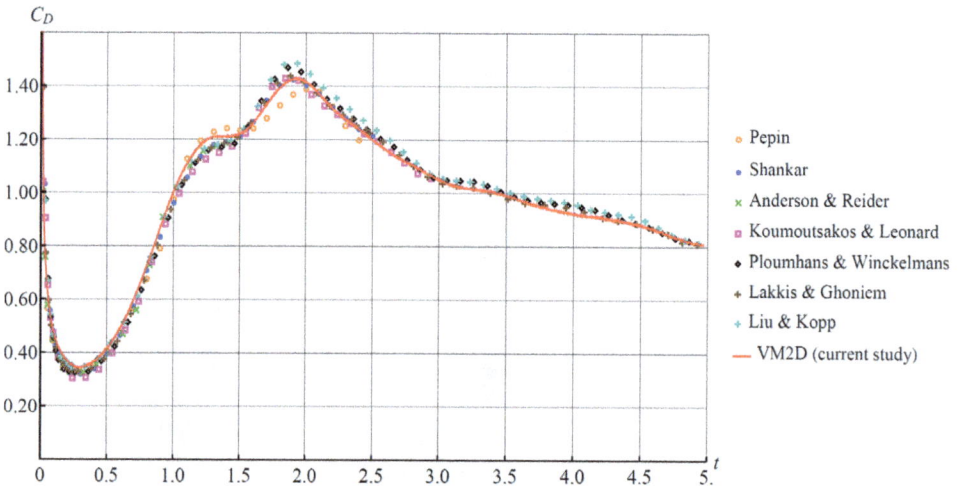

Figure 4:   Unsteady drag coefficient for flow simulation around impulsively starting cylinder at Re = 3,000.

4.2 Flow around rectangular airfoil

The set if problems of flow simulation around rectangular airfoils with different chord to thickness ratio (c/t) is considered. Example of vortex wake in quasi-steady regime behind the rectangular airfoil with $c/t = 7$ is shown in Fig. 5 at Re = 400, which is calculated with respect to the thickness t.

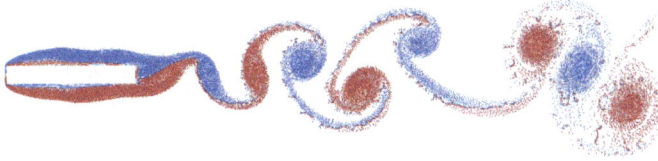

Figure 5: Vortex wake behind the rectangular airfoil.

For such set of problems an interesting effect can be observed. The Strouhal number calculated with respect to the airfoil chord is changing piecewise-constantly in dependence on the ratio c/t (Fig. 6). Numerical results are in a good agreement with results of other authors [24], [25].

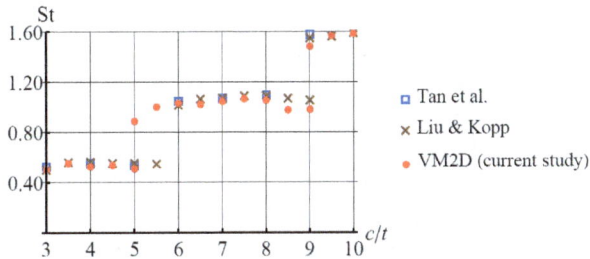

Figure 6: The Strouhal number against chord to thickness ratio.

4.3 Flow simulation around sphere

In VM3D for BIE solving the T-scheme with a piecewise-constant solution representation on triangular panels is implemented as well as a special algorithm of the solution correction to provide its divergence-free, based on the recovery of the double-layer potential density. The model problem of flow simulation around the sphere with unit diameter is considered. The surface is discretized by triangle panels of approximately equal areas. The steady regime of flow simulation is shown in Fig. 7.

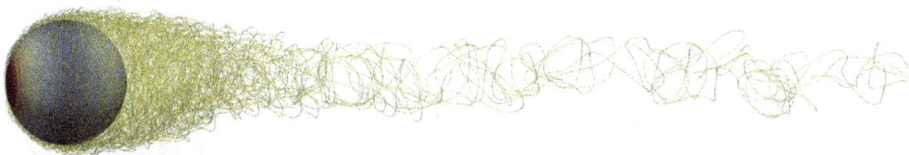

Figure 7: Vortex wake after the sphere.

Results of the computed averaged dimensionless pressure coefficient in dependence on the azimuthal angle of a point on the sphere are in good agreement with the experimental data as shown in Fig. 8.

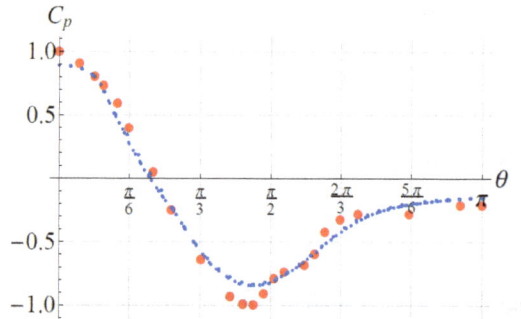

Figure 8:   Dimensionless pressure coefficient against on the azimuthal angle: numerical experiment (blue points) and experimental data (red points).

4.4 Added masses calculation

For FSI problems different coupling schemes are known. The simplest weakly-coupled approach is the splitting scheme: hydrodynamic step of flow simulation for the surface motion according to the known law, and mechanic step for the surface motion under known forces. Note that if we deal with light body or airfoil such an approach leads to numerical instability. To overcome this problem strongly-coupled monolithic approach can be used when parameters of the vorticity and surface velocity are computed during single system solving. This scheme is more complicated to implement and is not universal, so an iterative semi-implicit scheme [26] is preferable, which requires a preliminary calculation of the necessary components of the added mass tensor.

Components of the matrix of added masses can be expressed through the solution of the BIE (intensity of the vortex sheet on the surface) during the solution of the set of problems: impulsive start of the body/airfoil with unit velocity along all 2 or 3 axes and its impulsive rotation with a unit angular velocity around all the axes. In fact it is necessary to perform only one time step for each problem.

In the 2D case matrix of the added mass tensor consist of six components. Let's consider the problem of added mass computation for the Zhukovsky wing airfoil (Fig. 9) since the exact solution is known.

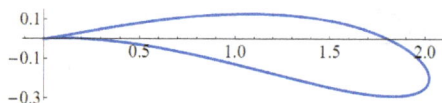

Figure 9: Zhukovsky airfoil shape.

Maximal relative errors are shown for all the components of the added mass tensor for a different number of panels are shown in Table 1, where three different numerical schemes are considered: N-scheme of the discrete vortex method and T-schemes with piecewise-constant (with index 0) and piecewise-linear (with index 1) solution representation.

Table 1: Relative errors of the added masses tensor estimation for the wing airfoil.

Number of panels	N-scheme	T^0-scheme	T^1-scheme
50	$3.4 \cdot 10^{-1}$	$5.7 \cdot 10^{-2}$	$1.7 \cdot 10^{-2}$
100	$6.2 \cdot 10^{-2}$	$2.5 \cdot 10^{-2}$	$4.6 \cdot 10^{-3}$
200	$3.9 \cdot 10^{-2}$	$1.3 \cdot 10^{-2}$	$1.7 \cdot 10^{-3}$
400	$2.2 \cdot 10^{-2}$	$6.9 \cdot 10^{-3}$	$7.4 \cdot 10^{-4}$
800	$1.2 \cdot 10^{-2}$	$3.7 \cdot 10^{-3}$	$3.5 \cdot 10^{-4}$
1,600	$6.1 \cdot 10^{-3}$	$2.0 \cdot 10^{-3}$	$1.7 \cdot 10^{-4}$
3,200	$3.4 \cdot 10^{-3}$	$1.1 \cdot 10^{-3}$	$8.5 \cdot 10^{-5}$

All the schemes provide close to the first order of accuracy. It is connected with exact solution behaviour in the x-directed motion the solution is bounded and has the first kind discontinuity at the cusp, however, for the y-directed motions the exact solution has a weak singularity at the cusp point. Values of added masses corresponding to the x-direction are calculated with the second order of accuracy when using the T^1-scheme. Note that in order to achieve an error level less than 1% for the N-scheme the airfoil should be discretized by 950 panels, while for T^0-scheme by 265 and for T^1-scheme by only 66 panels. For providing error level less than 0.1% it is required 10,800 panels for N-scheme, and 3,700 and only 307 for T-schemes respectively.

In 3D case, the problem of a three-axial ellipsoid is considered. It is known that for ellipsoid in principal axes, that are co-directional with its own axes, coefficients of added masses (λ_{11}, λ_{22}, λ_{33}) and coefficients of the added moments of inertia (λ_{44}, λ_{55}, λ_{66}) can be calculated exactly [27]. Relative errors of all components of the added mass tensor are given in Table 2 for ellipsoid with semiaxes $a = 0.5$, $b = 1.0$ and $c = 1.5$ discretized by triangle close to uniform mesh. For the BIE solving the T-scheme with piecewise-constant solution representation is used.

Table 2: Relative errors of added mass coefficients for ellipsoid.

Number of panels	$\delta\lambda_{11}$	$\delta\lambda_{22}$	$\delta\lambda_{33}$	$\delta\lambda_{44}$	$\delta\lambda_{55}$	$\delta\lambda_{66}$
186	0.1285	0.0713	0.0726	0.1081	0.2384	0.3200
488	0.0495	0.0275	0.0278	0.0420	0.1082	0.1427
954	0.0238	0.0151	0.0148	0.0248	0.0527	0.0699
2,060	0.0123	0.0064	0.0069	0.0093	0.0238	0.0338
3,780	0.0051	0.0045	0.0044	0.0066	0.0092	0.0126
7,780	0.0031	0.0017	0.0021	0.0017	0.0055	0.0088

It can be seen that the error decreases quite quickly with increasing of the number of panels; close to second order of accuracy is observed.

5 CONCLUSION

The present paper describes parallel open-source codes VM2D and VM3D are developed for the simulation of two- and three-dimensional viscous incompressible flows simulation around airfoils and bodies. T-schemes based on the Galerkin approach are developed and implemented for the numerical solution of the boundary integral equation. Test results show good agreement with the experimental data and other authors' results. At the same time, significantly lower level of non-stationary hydrodynamic loads oscillations can be observed in comparison with well-known N-schemes. The developed models and algorithms are suitable for numerical simulation in coupled FSI problems, including for light movable bodies/airfoils. Iterative semi-implicit approach is implemented, for which added masses tensor is calculated with rather high accuracy. Parallel implementation is performed using OpenMP, MPI and Nvidia CUDA technologies. The GPU-implementation is especially effective – one graphics accelerator Tesla V100 in terms of performance in VM2D can replace 84 28-core CPU nodes. There is also an opportunity to use several GPUs in parallel mode during one simulation, but it is not very efficient.

REFERENCES

[1] Kempka, S.N., Glass, M.W., Peery, J.S., Strickland, J.H. & Ingber, M.S., Accuracy considerations for implementing velocity boundary conditions in vorticity formulations. SANDIA Rep., SAND96-0583, 1996.

[2] Belotserkovsky, S.M. & Lifanov, I.K., *Method of Discrete Vortices*, CRC Press: New York, 464 pp., 1993.

[3] Marchevsky, I.K., Sokol, K.S. & Izmailova, I.A., *T*-schemes for mathematical simulation of vorticity generation on smooth airfoils in vortex methods. *Herald of the Bauman Moscow State Technical University, Series Natural Sciences*, 6(105), 2022. (In Russian.)

[4] Chorin, A.J., Numerical study of slightly viscous flow. *Journal of Fluid Mechanics*, **57**, pp. 785–796, 1973.

[5] Degond, P. & Mas-Gallic, S., The weighted particle method for convection-diffusion equations. I. The case of an isotropic viscosity. *Mathematics of Computation*, **53**(188), pp. 485–507, 1989.

[6] Shankar, S. & van Dommelen, L., A new diffusion procedure for vortex methods. *J. Comput. Phys.*, **127**, pp. 88–109, 1996.

[7] Dynnikova, G.Y., The Lagrangian approach to solving the time-dependent Navier–Stokes equations. *Doklady Physics.*, **49**(11), pp. 648–652, 2004.

[8] Mimeau, C. & Mortazavi, I., A review of vortex methods and their applications: From creation to recent advances. *Fluids*, **6**, 68, 2021.

[9] Lifanov, I.K., *Singular Integral Equations and Discrete Vortices*, VSP: Utrecht, 475 pp., 1996.

[10] Lewis, R.I., *Vortex Element Methods for Fluid Dynamic Analysis of Engineering Systems*, Cambridge University Press: Cambridge, UK, 566 pp., 1991.

[11] vvflow: CFD software for performing flow simulations with viscous vortex domains (VVD) method. https://github.com/vvflow. Accessed on: 29 Jun. 2023.

[12] Omega2D: Two-dimensional flow solver with GUI using vortex particle and boundary element methods. https://github.com/Applied-Scientific-Research/Omega2D. Accessed on: 29 Jun. 2023.

[13] Omega3D: 3D flowsolverwith GUI using vortex particle and boundary element methods. https://github.com/Applied-Scientific-Research/Omega3D. Accessed on: 29 Jun. 2023.

[14] FLOW Unsteady Aerodynamics Suite. https://github.com/byuflowlab/FLOWUnsteady. Accessed on: 29 Jun. 2023.

[15] Chair of Modelling and Simulation of Structures, Prof. Dr. Guido Morgenthal. https://www.uni-weimar.de/en/civil-engineering/chairs/modelling-and-simulation-of-structures/software/. Accessed on: 29 Jun. 2023.

[16] Ryatina, E. & Lagno A., The Barnes–Hut-type algorithm in 2D Lagrangian vortex particle methods. *J. Phys.: Conf. Ser.*, **1715**, 012069, 2021.

[17] Ryatina, E., Marchevsky, I. & Kolganova A., Boundary integral equation solving in vortex method using the Barnes–Hut/multipole algorithm. *Proceedings of the 2022 Ivannikov Ispras Open Conference (ISPRAS)*, pp. 74–80, 2022.

[18] Pepin, F.M., Simulation of the flow past an impulsively started cylinder using a discrete vortex method. PhD thesis, California Institute of Technology, 1990.

[19] Shankar, S., A new mesh-free vortex method. PhD thesis, FAMU-FSU College of Engineering, Florida, 1996.

[20] Anderson, C.R., Vorticity boundary conditions and boundary vorticity generation for two-dimensional viscous incompressible flows. *J. Comput. Phys.*, **80**, pp. 72–97, 1989.

[21] Koumoutsakos, P. & Leonard, A., High-resolution simulations of the flow around an impulsively started cylinder using vortex methods. *J. Fluid Mech.*, **296**, pp. 1–38, 1995.

[22] Ploumhans, P. & Winckelmans, G.S., Vortex methods for high-resolution simulations of viscous flow past bluff bodies of general geometry. *J. Comput. Phys.*, **165**(2), pp. 354–406, 2000.

[23] Lakkis, I. & Ghoniem, A., A high resolution spatially adaptive vortex method for separating flows. Part I: Two-dimensional domains. *J. Comput. Phys.*, **228**(2), pp. 491–515, 2009.

[24] Liu, Z. & Kopp, G.A., High-resolution vortex particle simulations of flows around rectangular cylinders. *Comp. Fluids.*, **40**, pp. 2–11, 2011.

[25] Tan, B.T., Thompson, M.C. & Hourigan, K., Simulated flow around long rectangular plates under cross flow perturbations. *Int. J. Fluid Dyn.*, **2**, 1, 1998.

[26] Dynnikova, G.Y., Added mass in a model of a viscous incompressible fluid. *Doklady Phys.*, **64**, 397–400, 2019.

[27] Korotkin, A.I., *Added Masses of Ship Structures*. Series: Fluid Mechanics and its Applications, Vol. 88, Springer, 391 pp., 2009.

Author index

www.ingramcontent.com/pod-product-compliance
Lightning Source LLC
Chambersburg PA
CBHW062003190326
41458CB00009B/2950